G
H
I
J
K
7
6
5
4
3
2
1
RCTIC OCEAN
North America
ATLANTIC OCEAN
PACIFIC OCEAN
South America
N OCEAN
LEGEND
Asia Continent name
Country border

OXFORD ATLAS+ FOR AUSTRALIAN SCHOOLS

F–2

HASS | STEM | Inquiry | Coding

Contents

Map skills

Mapping our Earth

We live on Earth. Earth is a planet in space.

Earth is round like a ball.

Earth

A globe is a model of Earth.

globe

A map of the world shows what a globe would look like if it was flattened out.

This is a map of the world.

An atlas is a book of maps.

What is a map?

A map is a drawing that shows an area seen from directly above.

This is an image of Australia seen from space.

This is a map of Australia.

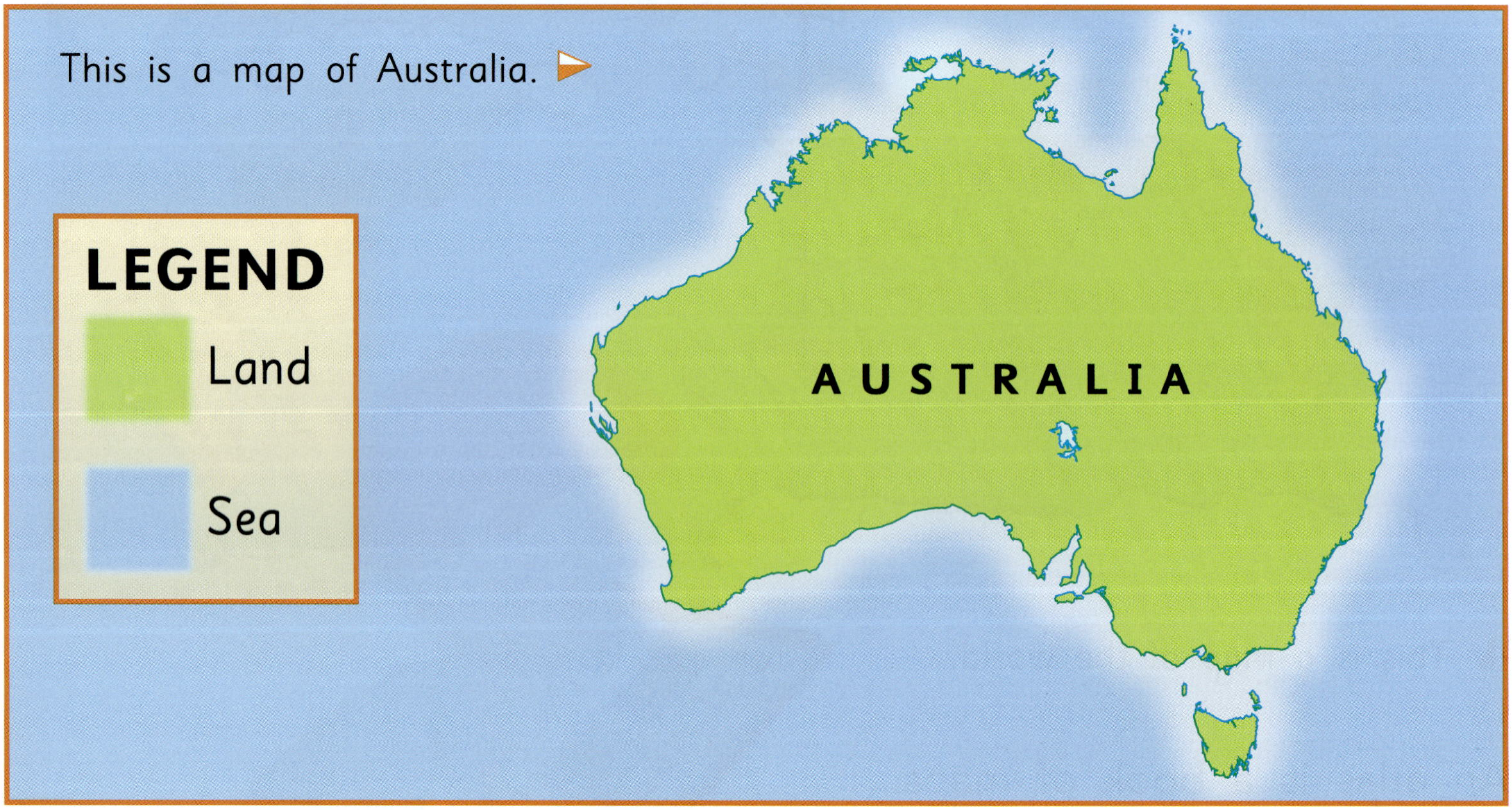

LEGEND

Land

Sea

AUSTRALIA

What is a legend?

A legend on a map shows symbols that represent real things in the world.

This map of Australia has symbols that represent deserts, mountains, rivers and a capital city.

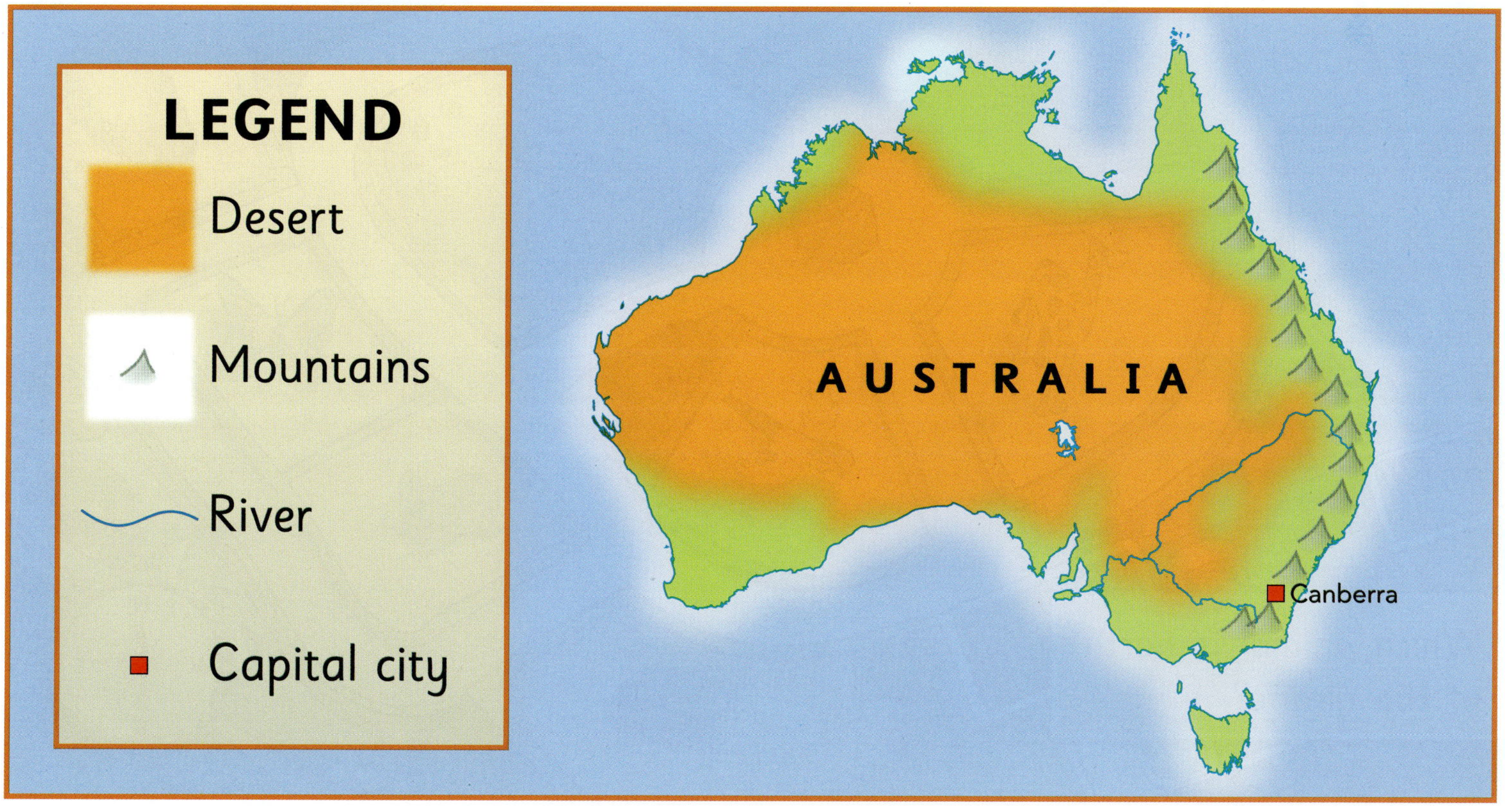

LEGEND

Desert

Mountains

River

Capital city

AUSTRALIA

Canberra

What is direction?

Direction is the way you go to get somewhere. A compass helps people to find out which direction they are going (north, south, east or west). The needle on a compass always points north.

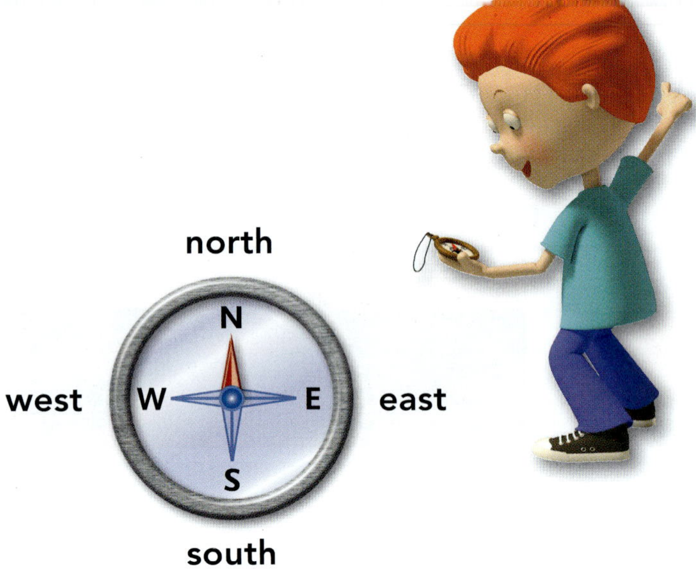

Some maps have a compass symbol on them to show where north is.

▶ Which animal is west of the turtle?

▶ Which animal is north of the giraffe?

What is a grid?

A grid divides a map into smaller areas. You can find each area by looking at the letters along the top or bottom of the map and the numbers down its sides.

B4 shows where the elephant enclosure is.

A B C D E F

5 4 3 2 1

Which grid square is the koala in?

What is scale?

I like going to the zoo with Mum. We get to the zoo by bus. There are lots of shops and buildings outside the zoo.

This is a view of the area outside the zoo.

The scale on a map helps people to work out how big the things shown on a map really are, and what the distance between them really is.

The scale is shown like this: Metres 0 10 20 30

This is a map of the area outside the zoo.

Introducing my world

Look around my places and my world. Learn more about maps and how to read them.

My room

This is my room. Look, I've just cleaned up and put my things away. Does your room look like this?

My bed is **next** to my desk. There is a lamp **above** my desk and there is a rubbish bin **below** my desk. What else can you see?

This is a view of Eddie's room.

This is a map of my room.

Can you find these things on the map?

N
W
E
S

Centimetres
0 50 100

This is a map of Eddie's room.

My house

I live with my mum, dad and my little sister, Pippa. Our house has lots of rooms. Some rooms are **big**, and some are **small**.

My room is **bigger** than Pippa's room but **smaller** than the lounge room. I can play **inside** the house or **outside** in the yard.

This is a view of Eddie's house.

This is a map of my house.

The map shows different parts of my house in different colours. The legend tells you which colour is used to show each part.

- Which room is shown in green?
- Can you find the biggest room in my house?
- How long is Pippa's bed?

This is a map of Eddie's house.

My street

I live in Smith Street. There are many houses in Smith Street. My **next-door** neighbours are Joe and Mr Chin.

If I turn **left** when I leave my house, I pass my friend Joe's house. Joe's house is **west** of my house.

This is a view of part of Eddie's street.

This is a map of part of my street.

If I turn **right** when I leave my house, I pass Mr Chin's house. Mr Chin's house is **east** of my house.

- What direction is the barbecue from my house?
- To cross Smith Street from my house, which direction do I need to go?

This is a map of part of Eddie's street.

My neighbourhood

I live **near** my school and not very **far** from the park. Sometimes, I walk to school with Dad.

We turn **left** onto the footpath. We turn **right** out of my street. Then we turn **right** at the roundabout.

This is a view of Eddie's neighbourhood.

This is a map of my neighbourhood.

On the map, my house is in grid square D2. My school is in D5.

- Which grid square is the playground in?
- What is in A2?
- What direction do I walk in to get home from school?

This is a map of Eddie's neighbourhood.

My city

I live in a **big** city. This is what it looks like from the **top** of a **tall** building. I can see my neighbourhood from here.

I can see the park in my neighbourhood. My house and my street look very **small** from here.

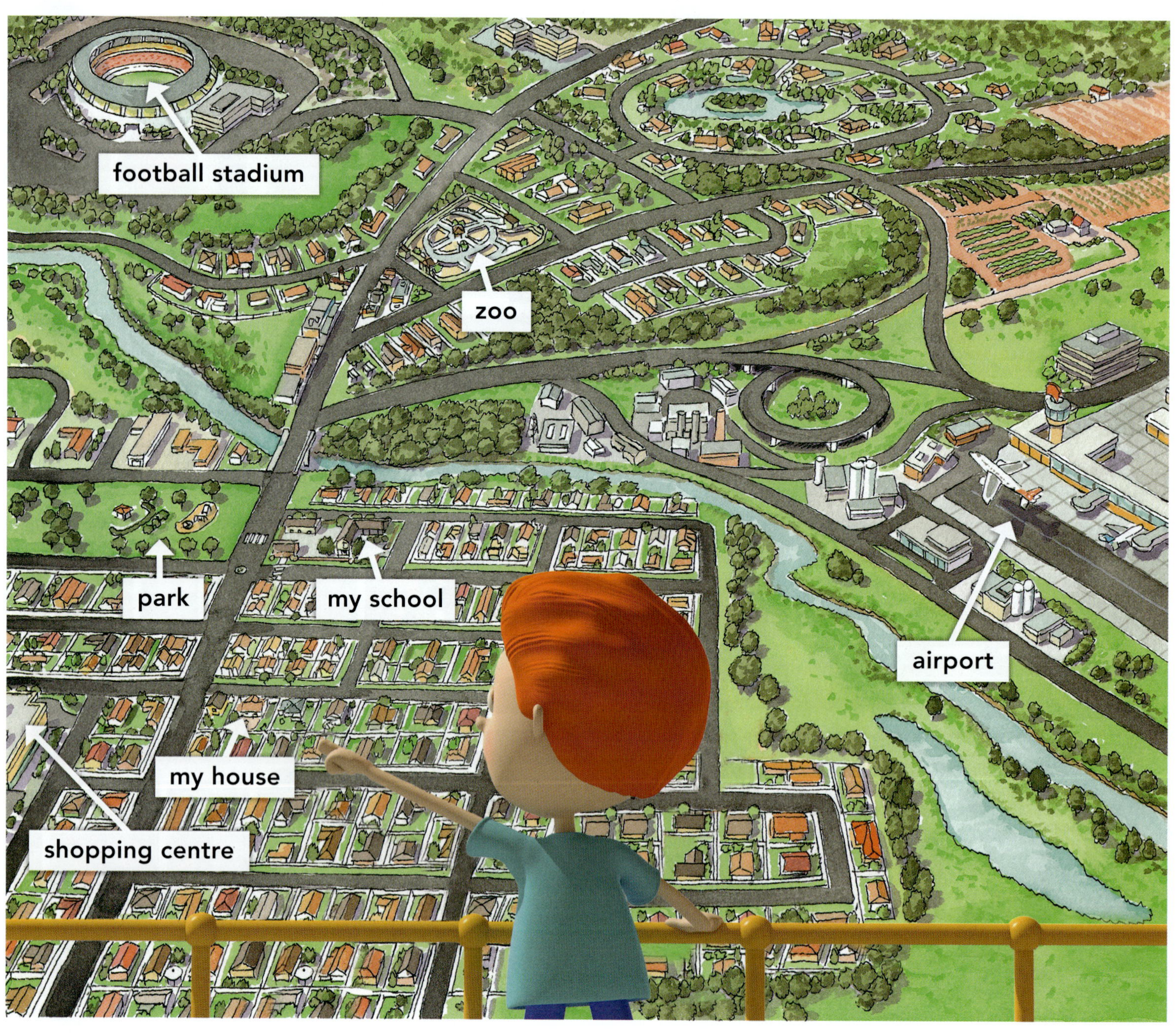

This is a view of Eddie's city.

This is a map of my city.

My house is **close** to Mills Park and my school but it is **far** away from the zoo.

- How far is Wilson Street from my house?
- What direction is Lake Franklin from my house?
- How long is Hill Lane?
- Which grid square is the airport in?

This is a map of Eddie's city.

Our world

Sun, Earth and Moon

The Sun is a huge star in space. Earth is a planet in space. Earth is made of rock. The Moon is made of rock, too.

Earth moves around, or orbits, the Sun. It takes one year, or 365 days, for Earth to orbit the Sun. The Moon orbits Earth.

The Sun is the Earth's main source of light and heat and, without it, living things could not survive. Without the light from the Sun, plants could not grow, and animals and humans would have no food to eat.

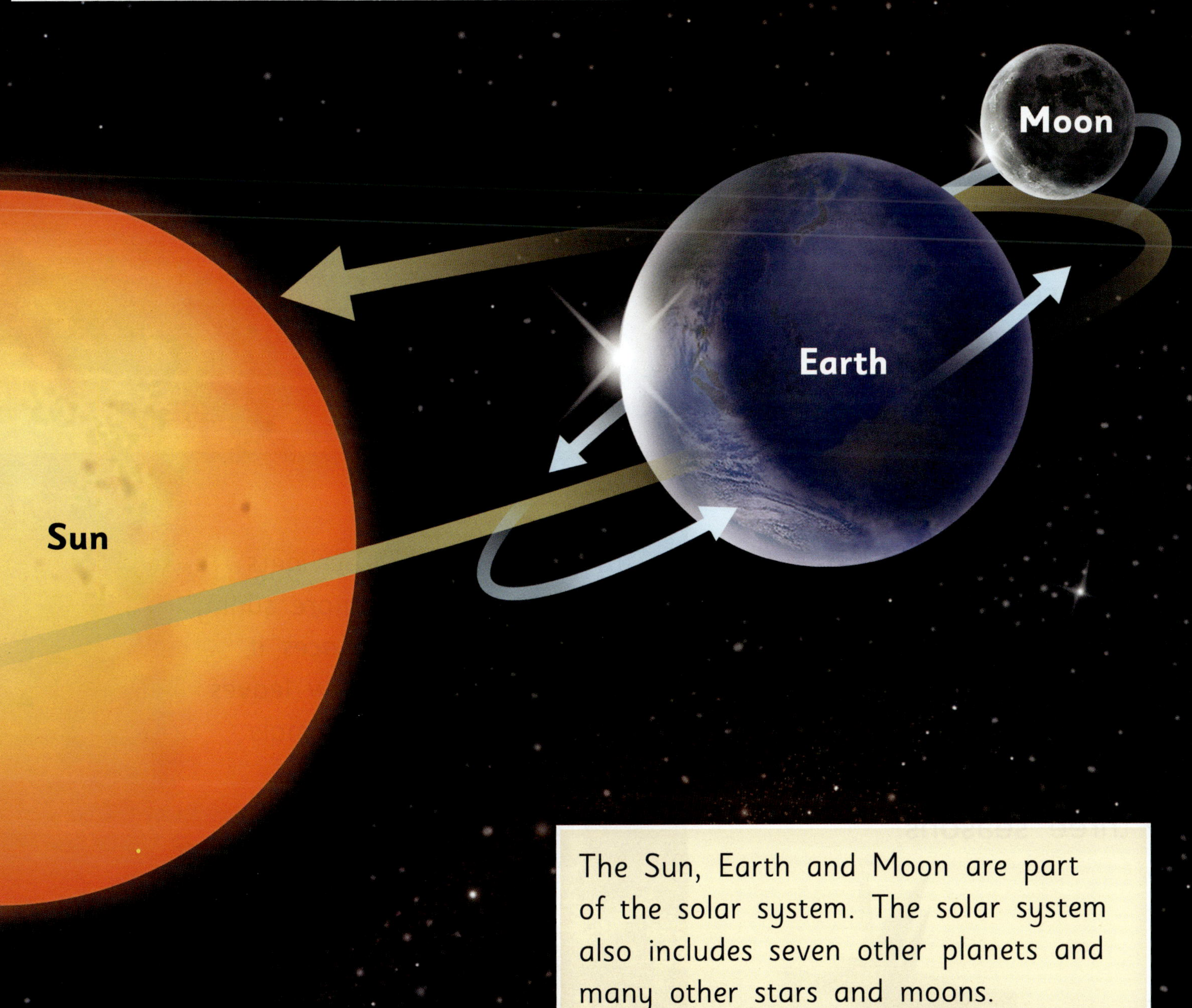

The Sun, Earth and Moon are part of the solar system. The solar system also includes seven other planets and many other stars and moons.

Seasons and weather

Many parts of Australia have four seasons: spring, summer, autumn and winter. Tropical parts of Australia have only two seasons: the wet season and the dry season.

Aboriginal and Torres Strait Island communities in different parts of Australia think about seasons in different ways. For the Nyoongar people, there are six seasons. For the Miriwoong people, there are three seasons.

Summer

Summer is the hottest season. In summer, people find ways to stay cool. Lots of people go swimming.

Spring

In spring, the weather gets warmer. There are more sunny days and in some places it rains. Plants grow new leaves and flowers. Many animals have babies.

Autumn

In autumn, the weather cools down. The leaves on some trees change colour and fall to the ground.

Winter

Winter is the coldest season. In winter, people find ways to stay warm. There is snow on the Australian Alps in winter.

Natural environments

Grasslands, mountains, ice and snow, deserts, coastlines and forests are natural environments around the world.

Environments are home to many different living things, including animals, plants and people.

Mountains

In Nepal, in Asia, more than half the country is made up of mountains. Many people live in these mountains.

Forests

For thousands of years, forests in Brazil have been home to many people, animals, insects and plants.

Ice and snow

Animals that live in the ice and snow often have thick fur to help keep them warm. Polar bears in the Arctic have thick fur to keep them warm.

Coastlines

Coastlines are where the land meets the sea. Australia's coastlines are home to many plants and animals.

Deserts

Although deserts are very dry, they are home to many plants, insects and animals. Some people live in deserts, too.

Grasslands

Grasslands in Africa are home to more animals than any other grasslands in the world.

Living things

Animals and plants are living things. They need air, food and water to live.

Some living things live on land. Some living things live in water. Some living things can even fly.

Plants use their roots to get water and food from soil.

Humans need water and food to survive.

Birds use their beaks to catch food and to feed their chicks.

Feathers keep birds warm.

Birds have sharp claws to catch food.

Fish have gills to breathe air in water.

Tigers have thick fur to keep warm in cold snow.

Tigers have sharp teeth to kill food and eat meat.

Tigers have sharp claws to catch and hold prey.

Life cycle of a frog

Animals have life cycles. They grow, change and have babies.

A frog begins life as an egg. A tadpole grows and changes into a frog during its life cycle.

1. Life begins

The female frog lays her eggs in water.

5. From froglet to adult frog

The froglet grows and loses its tail completely. It is now an adult frog.

2. Tadpole grows in egg

A tadpole grows in an egg. Its tail grows first, then its eyes, nostrils, gills and mouth grow.

3. Tadpole hatches

The tadpole hatches from its egg. It can swim underwater and breathe with its gills.

4. From tadpole to froglet

The tadpole changes into a young frog, called a froglet. It grows legs and arms. It loses its gills and grows lungs. Its tail starts to shrink. The froglet can leave the water and breathe with its lungs.

Endangered animals

Some animals are endangered. This means there are not many of them left in the wild and they are in danger of becoming extinct.

An animal becomes endangered when its environment changes or when it is hunted by people too much.

Green sea turtle

The green sea turtle lives in tropical seas around the world. Its environment is being destroyed by pollution. Green sea turtles and their eggs are also hunted for food.

Harpy eagle

The harpy eagle is one of the world's largest eagles. It lives in tropical rainforests in South America. People are destroying its environment for farms and logging.

Orang-utan

The orang-utan lives in the South-East Asian island of Sumatra. Its rainforest habitat is being destroyed to make way for palm oil plantations. Palm oil is used in many food items that we find in our supermarkets.

Leadbeater's possum

The Leadbeater's possum lives in forests in Australia. People are logging the trees that it lives in. Its environment has also been destroyed by bushfires.

Snow leopard

The snow leopard lives in the mountains of central and southern Asia. The snow leopard is hunted for its fur.

Red panda

The red panda lives in forests in the Himalayas. It eats bamboo, grasses, fruit and insects. Logging and farming are destroying its forest home.

Water on Earth

Water is Earth's most important resource. People, animals and plants cannot live without water.

Australia is a very dry country. Much of the natural environment in Australia is desert, so water is a very valuable resource. It is important that people don't waste water, so it doesn't run out.

Australia has many droughts. These are times when it does not rain much and water dries up.

Water stored in dams is cleaned, filtered and then pumped through underground pipes until it reaches our homes.

Animals need water to survive.

Crops need water to grow.

Dams are used to store water.

Many people use water tanks to collect rainwater.

Caring for the environment

Rubbish pollutes Earth. We need to recycle our rubbish to help care for Earth's environments.

Recycle

Recycle as many things as possible. Lots of rubbish from home can be recycled.

How is paper recycled?

1. A truck takes paper to a recycling plant.

2. The paper is sorted.

3. The paper is made into bales.

4. Then the bales are soaked in water to make pulp.

5. Rubbish is taken out of the pulp.

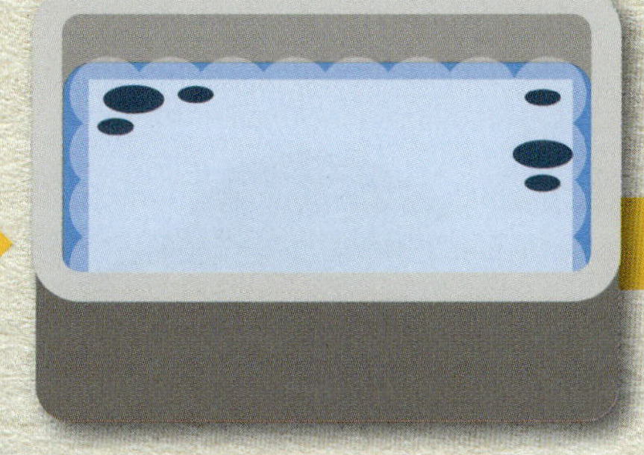

6. The pulp is cleaned.

7. The pulp is made into new paper.

Our amazing senses

We use our five senses of smell, taste, touch, sight and hearing to explore the world around us.

Our eyes help us to see shapes, colours, movement and light. The greatest source of light on Earth is the Sun. Light bulbs and lamps, candles, fire and the stars also give us light.

The skin on our body allows us to touch and feel the things around us. These pebbles feel smooth.

The taste buds on our tongue help us to taste foods. Foods can taste sweet, sour, bitter or salty. Watermelon tastes sweet.

We use our noses to smell. Some things smell good and others smell bad. I love the smell of lemons.

Our ears help us to hear sounds. Sounds are made when an object vibrates. When my dog barks, the air around my dog vibrates, too. The moving air will then cause my eardrum to vibrate and I can hear the sound.

I love playing music. I can make sounds on these instruments by doing different things. If I strum this guitar it will make lovely sounds.

I can shake these maracas to make music.

Push and pull

A push is a movement, or a force. A pull is a movement, or a force, too. We push and pull things to make them move.

Push
We push things to make them move.

This boy is pushing a toy car.

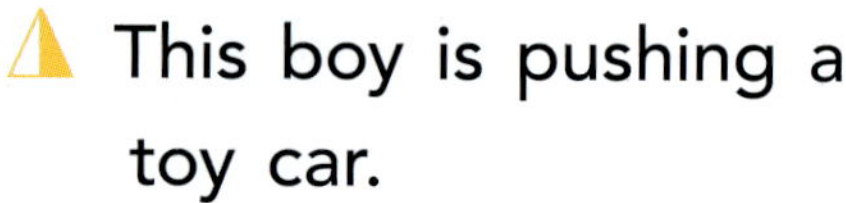

The shape and surface of an object change the way it moves when it is pushed or pulled. If we push a ball, it will move further than if we push a block of wood with the same force.

This boy is pushing his sister on the swing.

A strong push makes the toy car go faster.

Pull

We pull things to make them move.

This girl is pulling a toy.

This toy changes shape when it is pulled.

This horse is pulling a cart.

Places from the past

Places can be special for different reasons. Some places are special because they help us learn about our history. We need to look after these special places.

Dampier Archipelago, Western Australia
The Dampier Archipelago has some of the oldest rock art in the world. Rock art tells us about Aboriginal and Torres Strait Islander peoples' past.

Australian War Memorial, ACT

The Australian War Memorial is a special place in Canberra. It is a place where we can remember all the people who fought in wars.

Beechworth Post Office, Victoria

The Beechworth Post Office was built in 1870 during the gold rush. The gold rush is an important part of Australian history. Many places where gold was found grew into towns and cities.

Australia's first people

Aboriginal and Torres Strait Islander people are Indigenous Australians. They are Australia's first people. Aboriginal and Torres Strait Islander people have been living in Australia for at least 65 000 years.

Country

Aboriginal people believe that Australia was created in the Dreaming by Spirit Beings. Where a person lives, or where their ancestors lived, is called Country. Aboriginal people say people must care for Country.

Kinship

Aboriginal and Torres Strait Islander peoples' languages have kinship words to describe the people in their family. The same words can also describe people in the community who are not part of their family. Kinship words are signals that tell people how to treat one another.

In some Indigenous families, the kinship word for mother is *birrii*. The kinship word for aunt is also *birrii*.

Aboriginal and Torres Strait Islander people across Australia use different names for the seasons based on the cycles of plants, animals and the environment. For example, the table below shows the three main seasons described by the Miriwoong people of Western Australia.

Season name	Type of weather	Months
Nyinggiyi-mageny	wet weather	December, January, February, March
Warnka-mageny	cold weather	April, May, June, July, August
Barndenyirriny	hot weather	September, October, November

Aboriginal people are Australia's first people.

My family tree

My name is Ling. I was born in China. My family moved to Australia when I was a baby.

I live with my mum, my dad, my little sister and my grandparents.

A person who has the same parents as you is your sibling. Mei is my sibling. A female sibling is called a sister. A male sibling is called a brother.

Grandfather Yao
Grandmother Ping
I have one aunty, one uncle and two cousins. They live in China.
Dad
Mum
Uncle Shu
Aunty Min
Me (Ling)
Mei
Cousin Zi
Cousin Wei

Families

There are many different types of families. Big families, small families, blended families, single-parent families and many others. Each family is unique and special!

My name is Afrah and this is my mum. There are only two of us in my family. We do lots of great things together like riding to the city and going to the zoo!

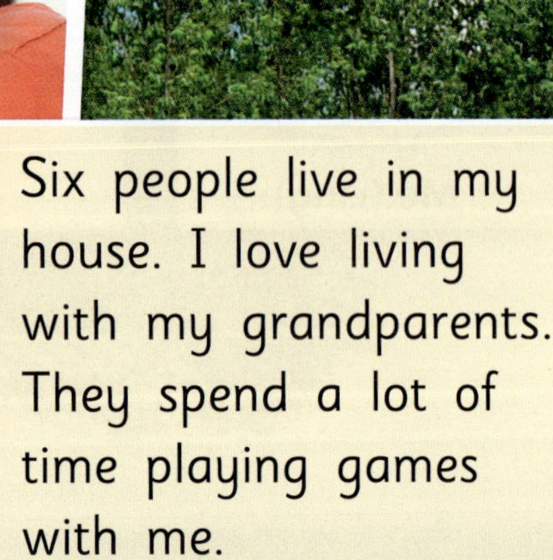

Six people live in my house. I love living with my grandparents. They spend a lot of time playing games with me.

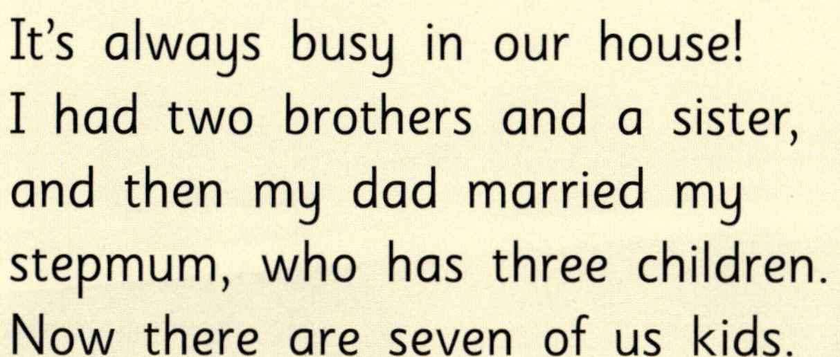

It's always busy in our house! I had two brothers and a sister, and then my dad married my stepmum, who has three children. Now there are seven of us kids.

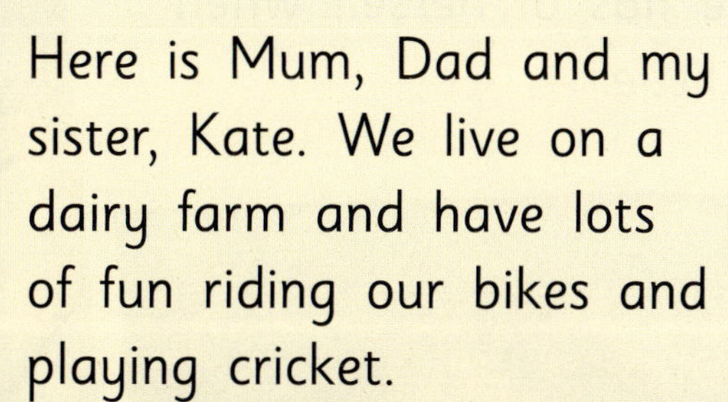

Here is Mum, Dad and my sister, Kate. We live on a dairy farm and have lots of fun riding our bikes and playing cricket.

I have lots of aunties and uncles, even if we are not all related. Here I am with my Uncle Jake, Aunty Nora and three of my cousins. We always have fun together.

Remembering the past

Photos, toys, songs and stories all help us to remember the past.

This is a photo of my grandmother in China when she was 10. This is the only photo she has of herself when she was little.

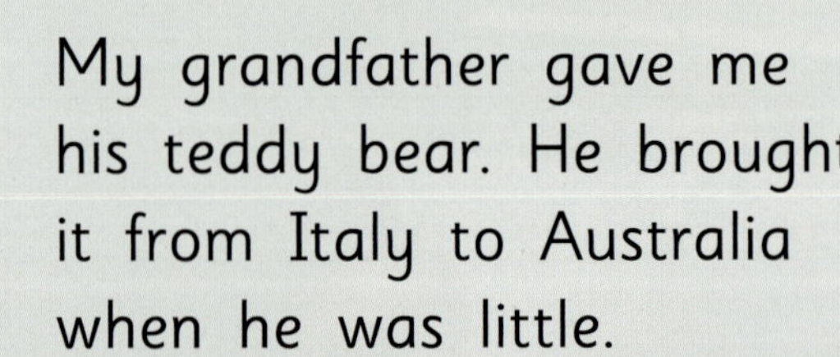

My grandfather gave me his teddy bear. He brought it from Italy to Australia when he was little.

My mum has a really big family with 11 brothers and sisters. My grandmother worked hard to look after the family. This is a quilt my grandmother made for my mum.

My grandfather loves playing his guitar. He sings songs about love, family and Country. His songs were passed down from his grandparents and their grandparents before that.

I love playing with this old phone – it's so big. My grandma lived on a farm with her family when she was a little girl. The telephone was the only way they could call other people. There were no mobile phones or computers and the postman only visited once a week.

Houses of different materials

Houses are safe places for people to live in. They shelter people from the weather. Houses are made of different materials in different places around the world.

Yurt, Mongolia

This house has a wooden frame. It has walls of felt made from sheep's wool. This house can be packed up and moved to a new place.

Stilt house, Brazil

This house is built on the bank of the Amazon River. The roof is made of leaves from the trees of the rainforest. This keeps the house dry when it rains. The house sits on stilts above the river.

Mudbrick house, Africa

This house is made from mudbricks. Mudbricks are made from clay and straw. They keep the house cool during the day and warm at night.

Glass house, United States of America

The walls of this house are made mostly of glass. Lots of sunlight shines into the house to keep it warm and bright.

Log cabin, Russia

This house is made of wooden logs. The logs came from pine trees in a nearby forest. Log cabins stay warm in winter and cool in summer.

Food from different cultures

Australia is a multicultural country. This means that people from many different cultures live in Australia.

People from different cultures have different kinds of food.

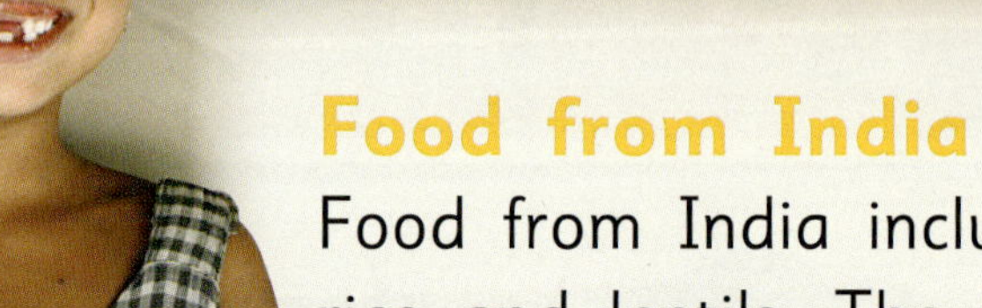

Food from India

Food from India includes rice and lentils. The food can be very spicy.

◄ This Indian girl is eating curry with rice.

▲ This girl is eating pasta.

▲ Many Mexican people buy their fruit and vegetables at the market.

Food from Italy

Pizza, pasta, salami and gelati are all Italian foods.

Food from Mexico

Food from Mexico includes beans and bread made from corn. Sauces made from tomatoes and chillies are popular, too.

Food from Indonesia

Food from Indonesia includes lots of seafood and tropical fruits.

This Indonesian family is cooking their food outside.

Food from China

Food from China includes rice, noodles, stir-fries and soups.

In China, people use chopsticks to eat their food.

Food from Sudan

Flatbread called *kisra* comes from Sudan. It is good to eat with stew.

This Sudanese woman is making *kisra*.

Celebrations around Australia

People all around Australia join in different celebrations and ceremonies. There are family celebrations such as birthdays, religious festivals such as Ramadan and Easter, and cultural festivals and ceremonies such as Harmony Day.

Each year, people celebrate Lunar New Year when the full moon appears between 21 January and 20 February. You can see lion and dragon dancers performing.

At many school and community events, there is an Acknowledgment of Country to show respect to the traditional owners of the land we live in.

Ramadan is a special month for Muslims. At the end of Ramadan, Muslims celebrate Eid al-Fitr. Families go to the mosque to pray, share food and give presents. They decorate their homes with beautiful coloured-glass lanterns.

NAIDOC Week is held in July every year to celebrate the culture and history of Aboriginal and Torres Strait Islander people. Schools around Australia celebrate by holding art exhibitions, creating posters and making bush tucker.

In Australia, Christmas is celebrated by many families. Some families go to church for Christmas Mass. People often put up Christmas trees, give each other presents and share a big meal.

Our country

Australia

Australia has six states and two territories. Its capital city is Canberra.

Parliament House in Canberra (H3)

- What is Australia's biggest state?
- What is the capital city of Queensland?
- Which capital city is found in G1?
- Which grid square is Melbourne found in?

There are more than 25 million people in Australia. Most people live in cities near the sea. The biggest city in Australia is Sydney.

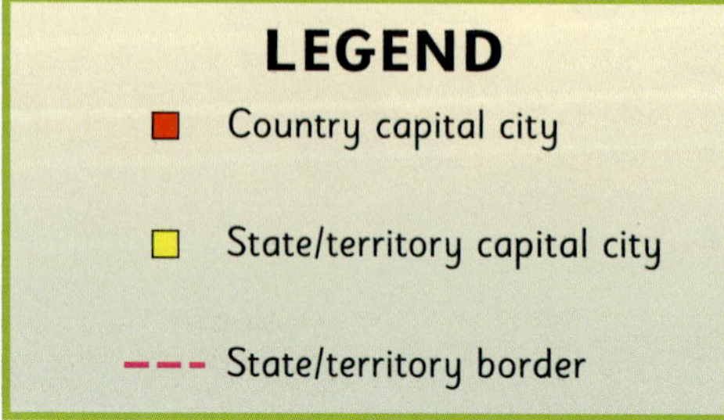

D
E
F
G
H
I
Torres Strait
Darwin
PACIFIC OCEAN
Gulf of Carpentaria
Northern Territory
Queensland
AUSTRALIA
Australia
Brisbane
South Australia
New South Wales
Great Australian Bight
Sydney
Adelaide
Canberra
Australian Capital Territory
Victoria
Melbourne
PACIFIC OCEAN
INDIAN OCEAN
Bass Strait
Tasmania
Hobart
8
7
6
5
4
3
2
1

About Australia

Australia is a country in the southern part of the world. It is a very dry continent. Much of Australia is desert.

Australia has many special plants and animals. It is home to animals called marsupials. Marsupials are animals with pouches, such as kangaroos and bilbies. Bilbies live in deserts. They eat insects, grubs and seeds.

Bilbies are marsupials.

- Which lake is found in South Australia?
- Which state capital city is found in B3?
- What type of land covers most of Australia?

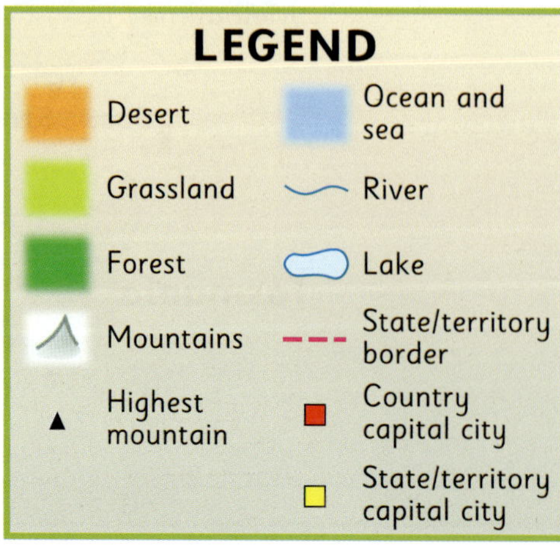

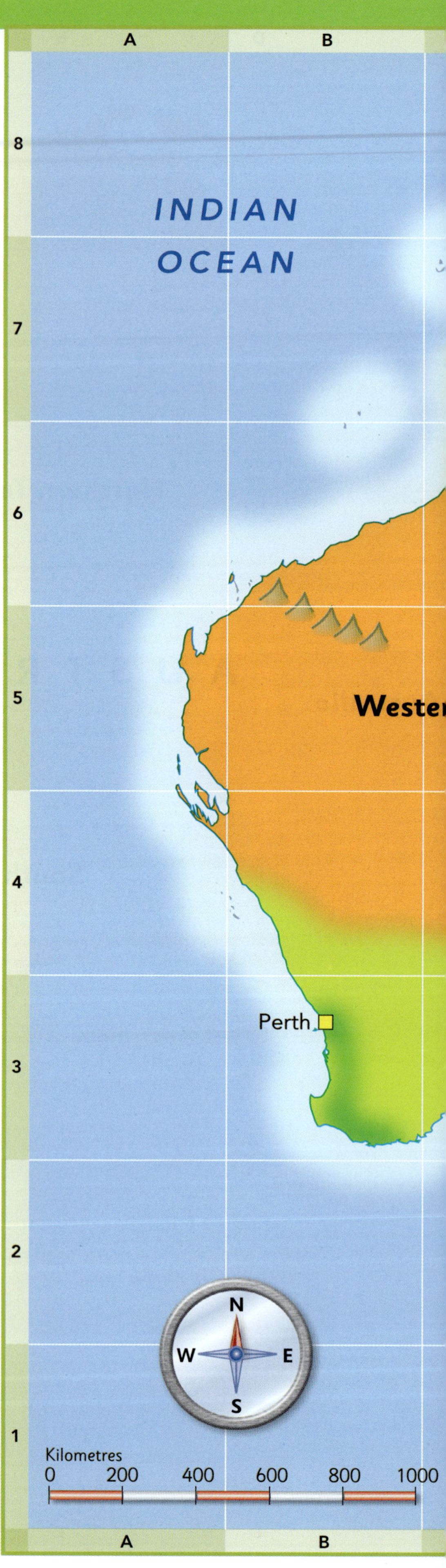

D
E
F
G
H
I
8
7
6
5
4
3
2
1
Torres Strait
ARAFURA SEA
Darwin
Gulf of Carpentaria
PACIFIC OCEAN
CORAL SEA
Northern Territory
Queensland
GREAT DIVIDING RANGE
AUSTRALIA
Western Australia
Great Victoria Desert
Kati Thanda-Lake Eyre
Brisbane
South Australia
Darling River
New South Wales
Great Australian Bight
Adelaide
Sydney
Canberra
Murray River
Australian Capital Territory
Mt Kosciuszko
Victoria
Melbourne
INDIAN OCEAN
PACIFIC OCEAN
Bass Strait
Tasmania
Hobart
TASMAN SEA

Western Australia

Western Australia is Australia's biggest state. Its capital city is Perth.

There are forests in the south-west of Western Australia. Tall trees, called Karri trees, grow in these forests.

Three of Australia's four biggest deserts are in Western Australia. Some people in Western Australia work in the desert in the mining industry. Some Indigenous Australians live in desert communities. Many others live in towns and cities.

Karri trees are one of the tallest kind of tree in the world.

There are many desert communities in Western Australia.

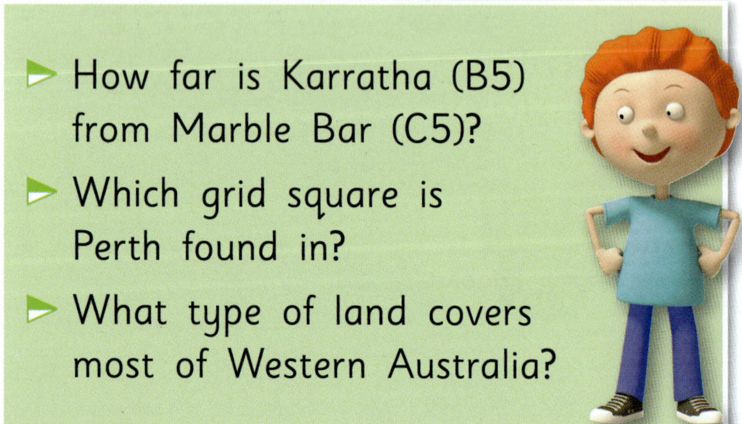

- How far is Karratha (B5) from Marble Bar (C5)?
- Which grid square is Perth found in?
- What type of land covers most of Western Australia?

A
B
C
D
E
F
7
6
5
4
3
2
1
N
W
E
S
TIMOR SEA
Kalumburu
Wyndham
Cape Leveque
Kilometres
0 100 200 300 400 500
Kimberley
Broome
Fitzroy River
Halls Creek
INDIAN
OCEAN
Dampier Archipelago
Port Hedland
Karratha
Great Sandy
Desert
Marble Bar
Exmouth
HAMERSLEY RANGE
Tom Price
Mt Meharry
Newman
Lake Mackay
Northern Territory
Gibson Desert
Western Australia
Carnarvon
Gascoyne River
Shark Bay
Murchison River
Wiluna
Kalbarri
Mount Magnet
Great Victoria
Desert
South Australia
Geraldton
Kalgoorlie
Rawlinna
Nullarbor Plain
Eucla
Perth
Rockingham
Mandurah
Norseman
Great Australian Bight
Bunbury
Busselton
Katanning
Esperance
Manjimup
Cape Leeuwin
Albany

Northern Territory

The Northern Territory runs from the centre of Australia to the north coast of Australia. Its capital city is Darwin.

Many people in the Northern Territory work in the mining and tourism industries.

Kakadu National Park is popular with tourists. It is in the north of the territory. Many tourists also visit Uluṟu, in the south of the territory. Both Kakadu and Uluṟu are very special to Aboriginal people.

Uluṟu, in Uluṟu-Kata Tjuṯa National Park (C1), is a World Heritage Site.

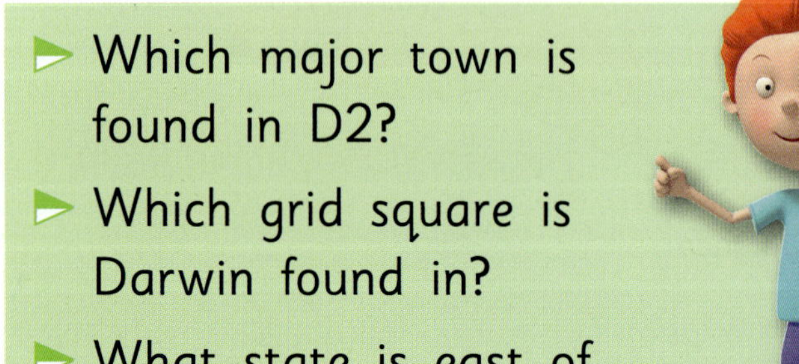

- Which major town is found in D2?
- Which grid square is Darwin found in?
- What state is east of the Northern Territory?

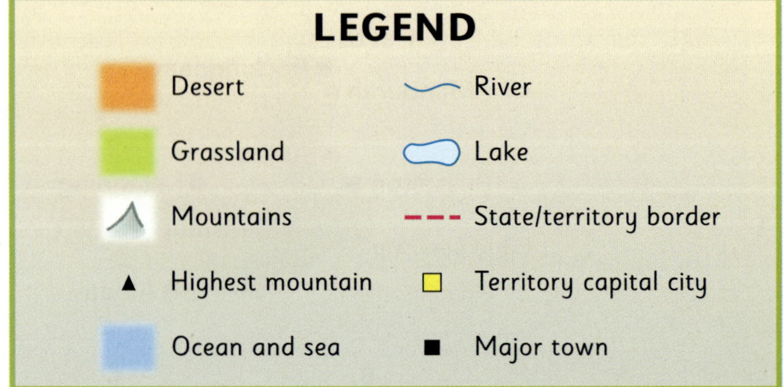

A
B
C
D
E
F
ARAFURA SEA
N
W
E
S
Melville Island
Van Diemen Gulf
Warruwi
Maningrida
Nhulunbuy
Darwin
Palmerston
Oenpelli
Kakadu National Park
Kilometres
0 50 100 150 200 250
Adelaide River
Daly River
Joseph Bonaparte Gulf
Wadeye
Katherine
Groote Eylandt
Numbulwar
Roper River
Gulf of Carpentaria
Victoria River
Timber Creek
Borroloola
Daly Waters
Tanami Desert
Tanami
Tennant Creek
Western Australia
Northern Territory
Queensland
Barrow Creek
Mt Zeil
Alice Springs
Lake Amadeus
Uluru-Kata Tjuta National Park
Finke
Simpson Desert
7
6
5
4
3
2
1

South Australia

The capital city of South Australia is Adelaide.

Farming is important in the southern part of the state. Many kinds of crops are grown, such as wheat and grapes.

A lot of mining takes place in South Australia. Opals are mined at Coober Pedy.

Some people in Coober Pedy live under the ground. Living underground helps them to keep cool in the desert.

Grapes grow on vines.

Underground house in Coober Pedy (C5)

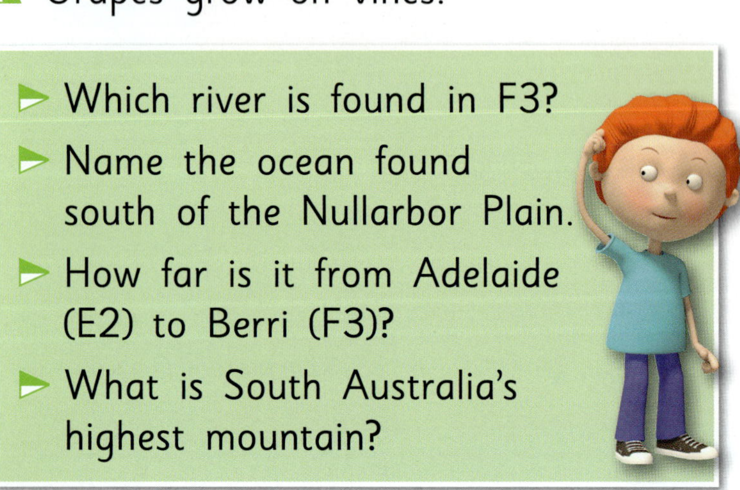

- Which river is found in F3?
- Name the ocean found south of the Nullarbor Plain.
- How far is it from Adelaide (E2) to Berri (F3)?
- What is South Australia's highest mountain?

A
B
C
D
E
F
7
6
5
4
3
2
1
Northern Territory
Queensland
Mt Woodroffe
Great Victoria
Desert
Oodnadatta
Cooper Creek
Kati Thanda-Lake Eyre
Coober Pedy
Western Australia
Marree
South Australia
Leigh Creek
Andamooka
Ooldea
Hughes
Tarcoola
FLINDERS RANGES
New South Wales
Nullarbor Plain
Yalata
Ceduna
Port Augusta
Streaky Bay
Great Australian Bight
Whyalla
Port Pirie
Cowell
Kadina
River
Murray
Spencer
Gulf
Berri
Gawler
Port Lincoln
Gulf
St
Vincent
Adelaide
Yorketown
Murray Bridge
INDIAN
OCEAN
Kangaroo
Island
Victoria
N
W
E
S
Kingston
Naracoorte
Mount Gambier
Kilometres
0 50 100 150 200 250

Queensland

Queensland is in the north-eastern part of Australia. Its capital city is Brisbane.

Many people in Queensland work in the farming, mining and tourism industries.

Thousands of tourists visit Queensland every year. They go to enjoy the beaches, rainforests and islands.

Tourists also visit World Heritage Sites, such as the Great Barrier Reef and the Daintree Rainforest.

Great Barrier Reef (C6)

Daintree Rainforest (C5)

- Which island is north of Cape York Peninsula?
- What major town is west of Hughenden?
- Which grid square is the capital city of Queensland found in?

A
B
C
D
E
F
7
6
5
4
3
2
1
Torres Strait
Thursday Island
CORAL SEA
N
W
E
S
Gulf of Carpentaria
Cape York Peninsula
GREAT BARRIER REEF
Kilometres
0
100
200
300
400
500
Daintree National Park
PACIFIC OCEAN
Mornington Island
Cairns
Mt Bartle Frere
Normanton
Townsville
Flinders River
Burdekin River
GREAT DIVIDING RANGE
Lake Dalrymple
Mount Isa
Hughenden
Mackay
Queensland
Yeppoon
Rockhampton
Longreach
Emerald
Gladstone
Fraser Island
Simpson Desert
Bundaberg
Hervey Bay
Birdsville
Burnett River
Charleville
Gympie
Sunshine Coast
South Australia
Toowoomba
Brisbane
Ipswich
Warwick
Gold Coast

New South Wales

New South Wales is home to more people than any other state or territory in Australia. Over eight million people live in New South Wales. Its capital city is Sydney.

Australia's highest mountain is Mount Kosciuszko. It is a part of a long line of mountains called the Great Dividing Range. This mountain range runs through New South Wales, into Victoria and Queensland.

Mount Kosciuszko (E2)

- Find Australia's longest river at D2.
- Which Australian territory is found within the area of New South Wales?

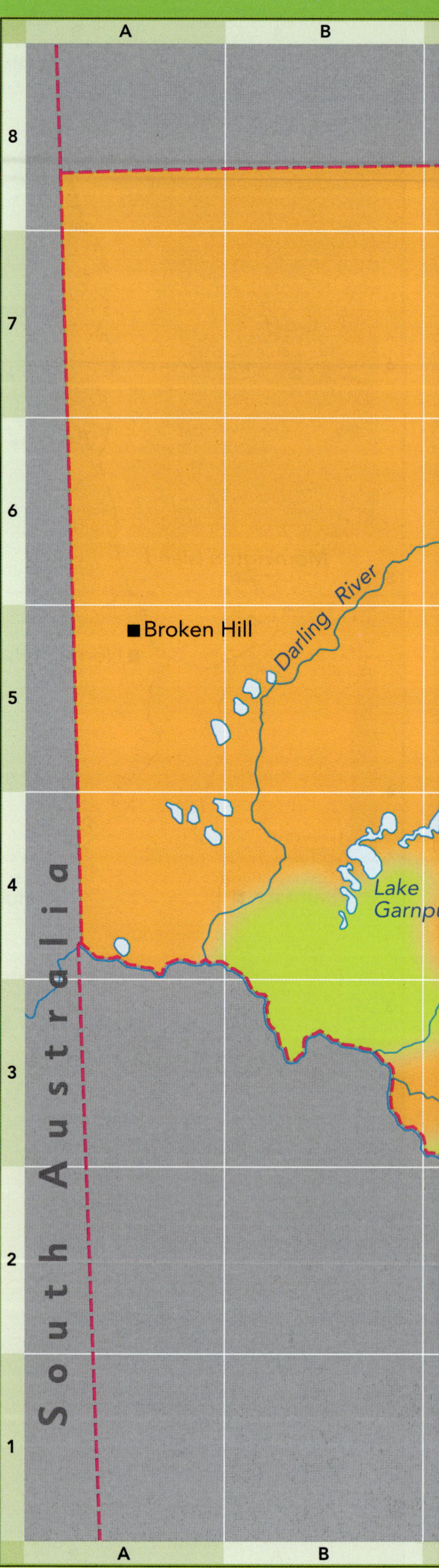

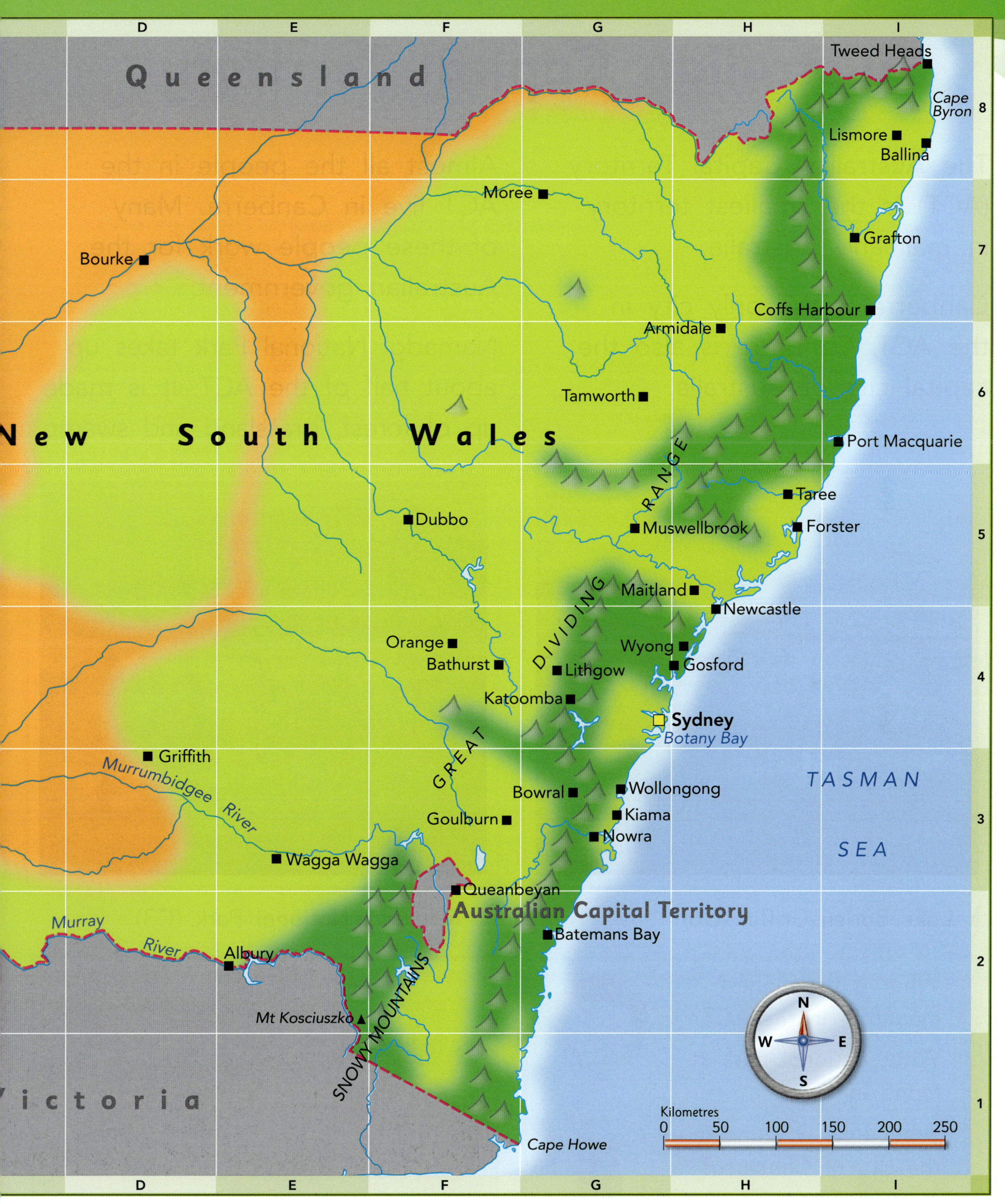

D
E
F
G
H
I
Queensland
New South Wales
Victoria
Tweed Heads
Cape Byron
Lismore
Ballina
Moree
Grafton
Bourke
Coffs Harbour
Armidale
Tamworth
Port Macquarie
RANGE
Taree
Dubbo
Muswellbrook
Forster
DIVIDING
Maitland
Newcastle
Orange
Wyong
Bathurst
Lithgow
Gosford
Katoomba
Sydney
Botany Bay
GREAT
Griffith
Murrumbidgee River
TASMAN SEA
Bowral
Wollongong
Kiama
Goulburn
Nowra
Wagga Wagga
Queanbeyan
Australian Capital Territory
Batemans Bay
Murray River
Albury
SNOWY MOUNTAINS
Mt Kosciuszko
N
W
E
S
Kilometres
0 50 100 150 200 250
Cape Howe
8
7
6
5
4
3
2
1

Australian Capital Territory

The Australian Capital Territory (ACT) is the smallest territory in mainland Australia.

Canberra is the only city in the ACT. Canberra is also the capital city of Australia.

Almost all the people in the ACT live in Canberra. Many of these people work for the Australian government.

Namadgi National Park takes up about half of the ACT. It is made up of forest, grassland and swamp.

Lake Burley Griffin (D6)

Namadgi National Park (C3)

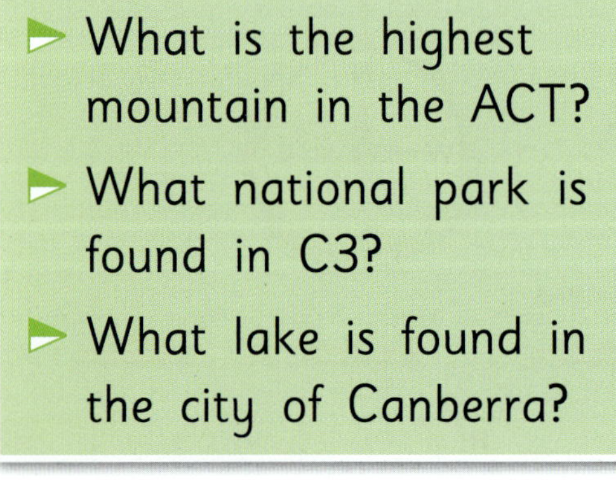

- What is the highest mountain in the ACT?
- What national park is found in C3?
- What lake is found in the city of Canberra?

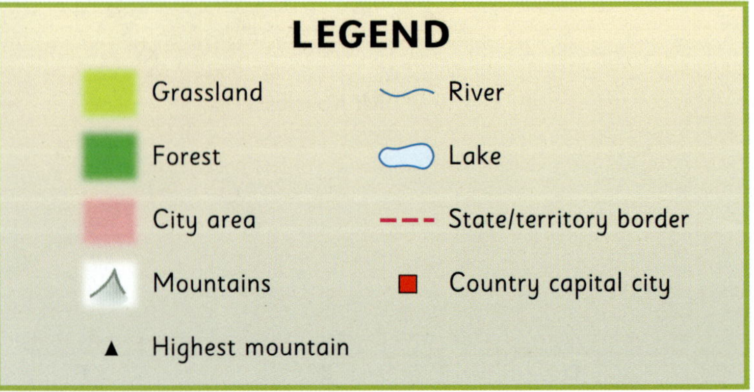

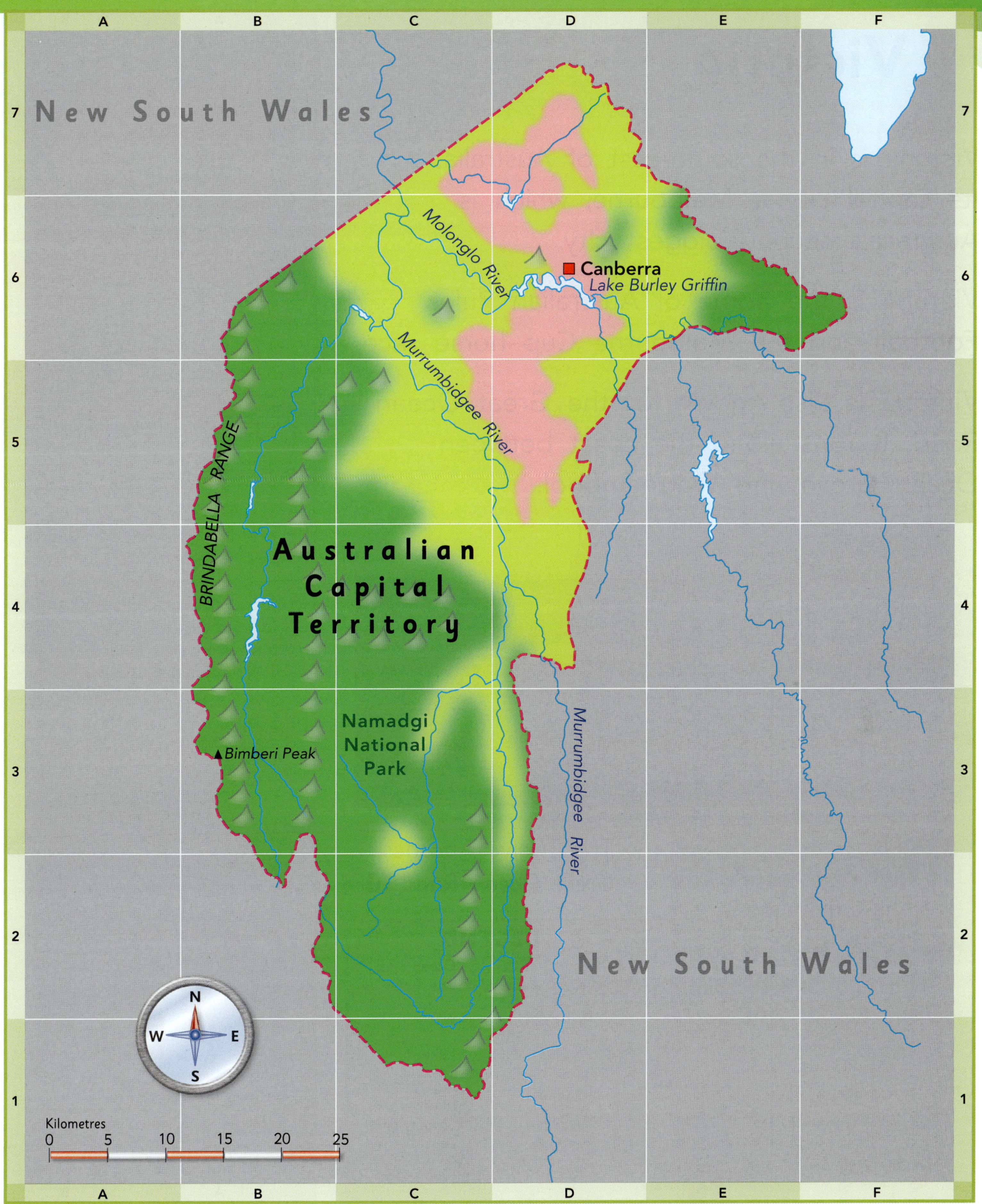
A
B
C
D
E
F
7
6
5
4
3
2
1
New South Wales
Molonglo River
Canberra
Lake Burley Griffin
Murrumbidgee River
BRINDABELLA RANGE
Australian Capital Territory
Namadgi National Park
Bimberi Peak
Murrumbidgee River
New South Wales
N
W
E
S
Kilometres
0
5
10
15
20
25

Victoria

Victoria is in the south-east of Australia. Its capital city is Melbourne, which is Australia's second biggest city.

Victoria is the home of Australian Rules Football and the Melbourne Cup horse race.

Victoria is also known for the Great Ocean Road. It runs along the coast between Ocean Grove and Warrnambool.

The Twelve Apostles, Great Ocean Road (C3)

- What is Victoria's highest mountain?
- How far is Warrnambool (B3) from Ocean Grove (D3)?
- What bay is Melbourne (E4) built around?

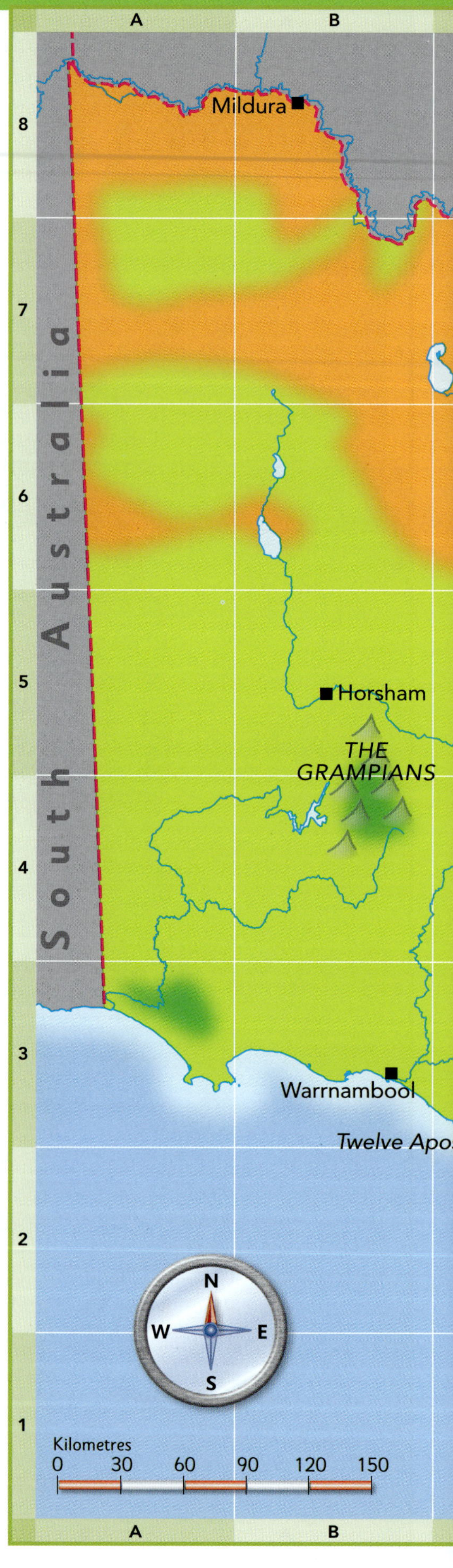

D
E
F
G
H
I
8
7
6
5
4
3
2
1
Murrumbidgee River
New South Wales
Australian Capital Territory
Murray River
Echuca
Wodonga
Shepparton
Wangaratta
Victoria
Bendigo
Mt Bogong
Goulburn River
GREAT DIVIDING RANGE
SNOWY MOUNTAINS
Snowy River
Ballarat
Sunbury
Bacchus Marsh
Melbourne
Bairnsdale
Lake
rangamite
Geelong
Port Phillip Bay
Warragul
Moe
Traralgon
Sale
Morwell
Ocean Grove
Phillip Island
Wilsons Promontory
Bass Strait
Tasmania

Tasmania

Tasmania is the smallest state in Australia. It is made up of one main island. The capital city of Tasmania is Hobart.

Tasmania is colder and wetter than the rest of Australia.

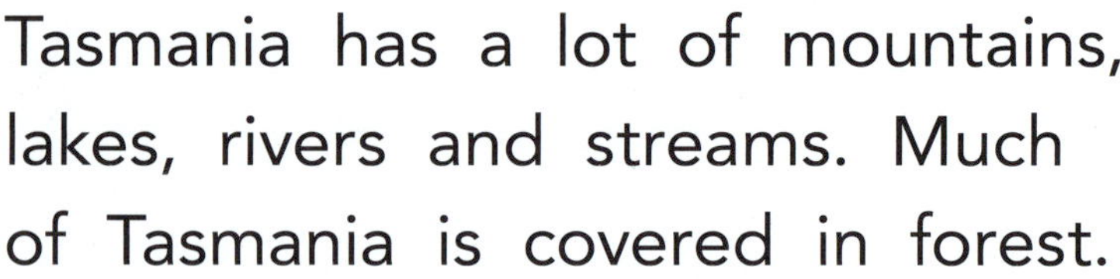

Tasmania has a lot of mountains, lakes, rivers and streams. Much of Tasmania is covered in forest.

Many people in Tasmania work on farms or in the timber, mining and fishing industries.

Cradle Mountain (C4) is popular with tourists.

Hobart (E2) is found near the mouth of the Derwent River.

- What major town is found in grid square E4?
- How far is Oatlands (E3) from Port Arthur (E2)?
- What is the name of the sea north of Tasmania?

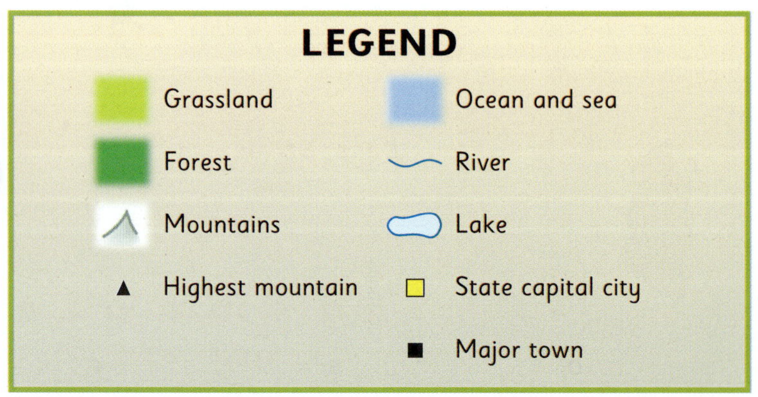

A
B
C
D
E
F
7
6
5
4
3
2
1
King Island
Bass Strait
Flinders Island
Cape Barren Island
Cape Grim
Stanley
Smithton
Bridport
Burnie
Devonport
Tamar River
Derby
St Helens
Savage River
Launceston
Deloraine
South Esk River
Fingal
St Marys
Cradle Mountain
INDIAN OCEAN
Rosebery
Mt Ossa
Tasmania
Queenstown
Strahan
Derwent Bridge
Oatlands
Gordon River
Derwent River
Strathgordon
Lake Gordon
Hobart
Port Arthur
Tasman Peninsula
N
W
E
S
Kilometres
0 20 40 60 80 100
South West Cape
South East Cape
TASMAN SEA

The world

This map of the world shows the continents, countries and oceans. There are seven continents.

Most continents are made up of many different countries.

There are more than 190 countries in the world.

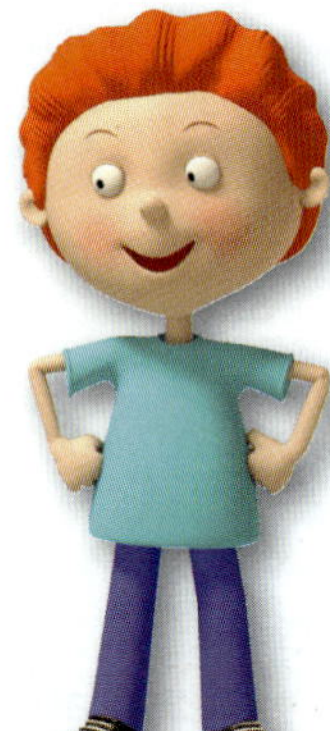

- What is the smallest continent?
- Which continent is found in the Southern Ocean?

LEGEND

Asia Continent name

Country border

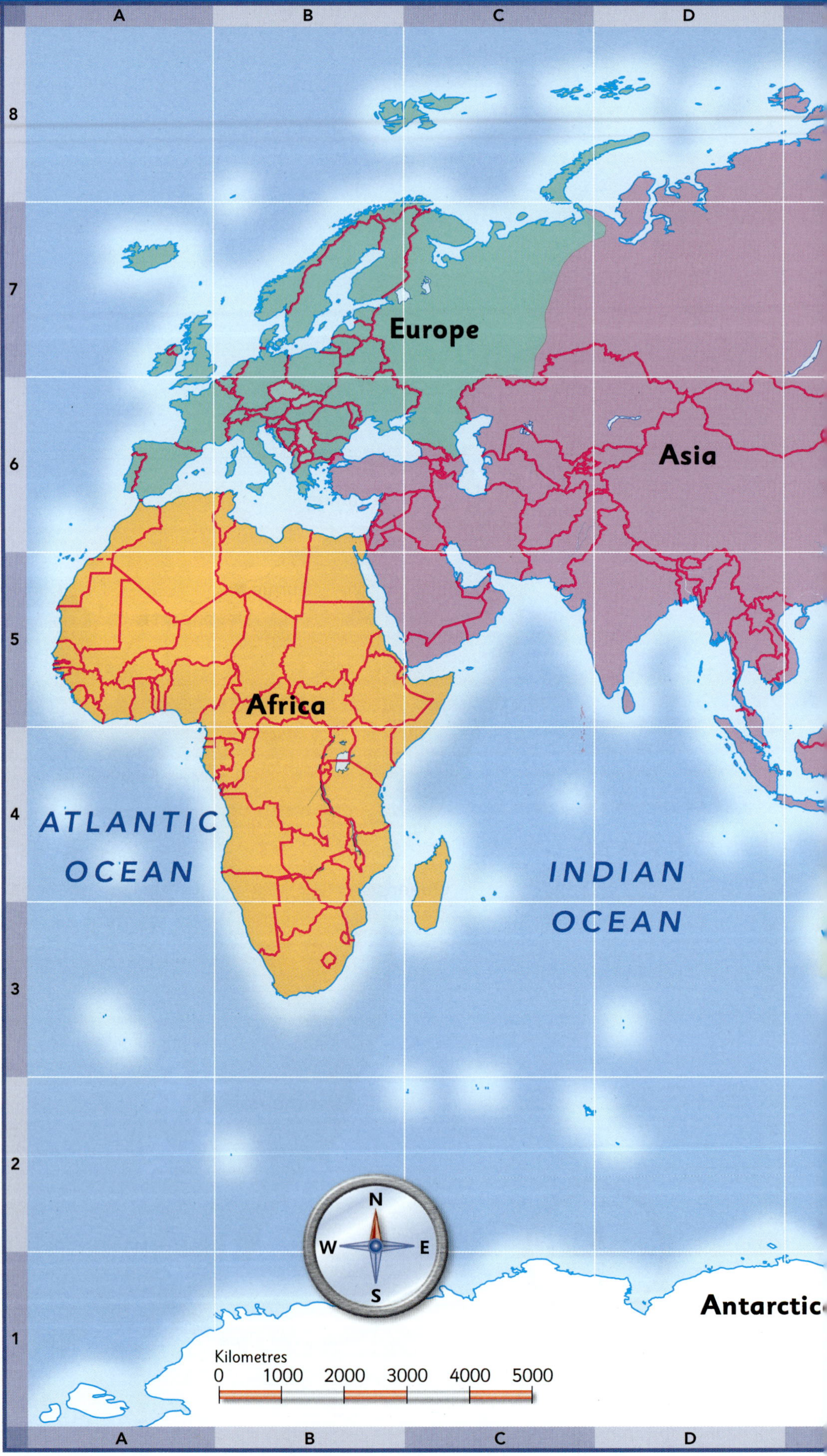

A
B
C
D
8
7
6
5
4
3
2
1
Europe
Asia
Africa
ATLANTIC OCEAN
INDIAN OCEAN
Antarctic
N
W
E
S
Kilometres
0
1000
2000
3000
4000
5000

F
G
H
I
J
K
ARCTIC OCEAN
North America
ATLANTIC OCEAN
PACIFIC OCEAN
South America
stralia
SOUTHERN OCEAN
8
7
6
5
4
3
2
1

The Pacific

There are 17 countries in the Pacific area. These include Australia, New Zealand, Papua New Guinea and 14 small island countries.

Australia is the only continent in the Pacific area. The Pacific Ocean is the biggest ocean on Earth.

Joe is from Papua New Guinea (B5). He is paddling a canoe.

- What is the capital city of Fiji (D4)?
- Which grid square is Nauru found in?
- Which island is east of Pitcairn Island?

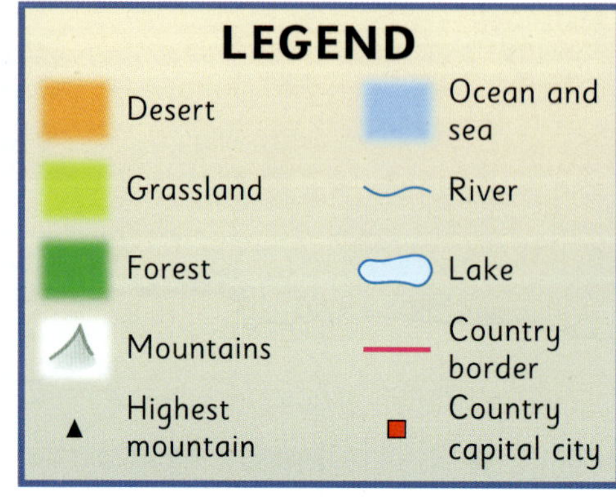

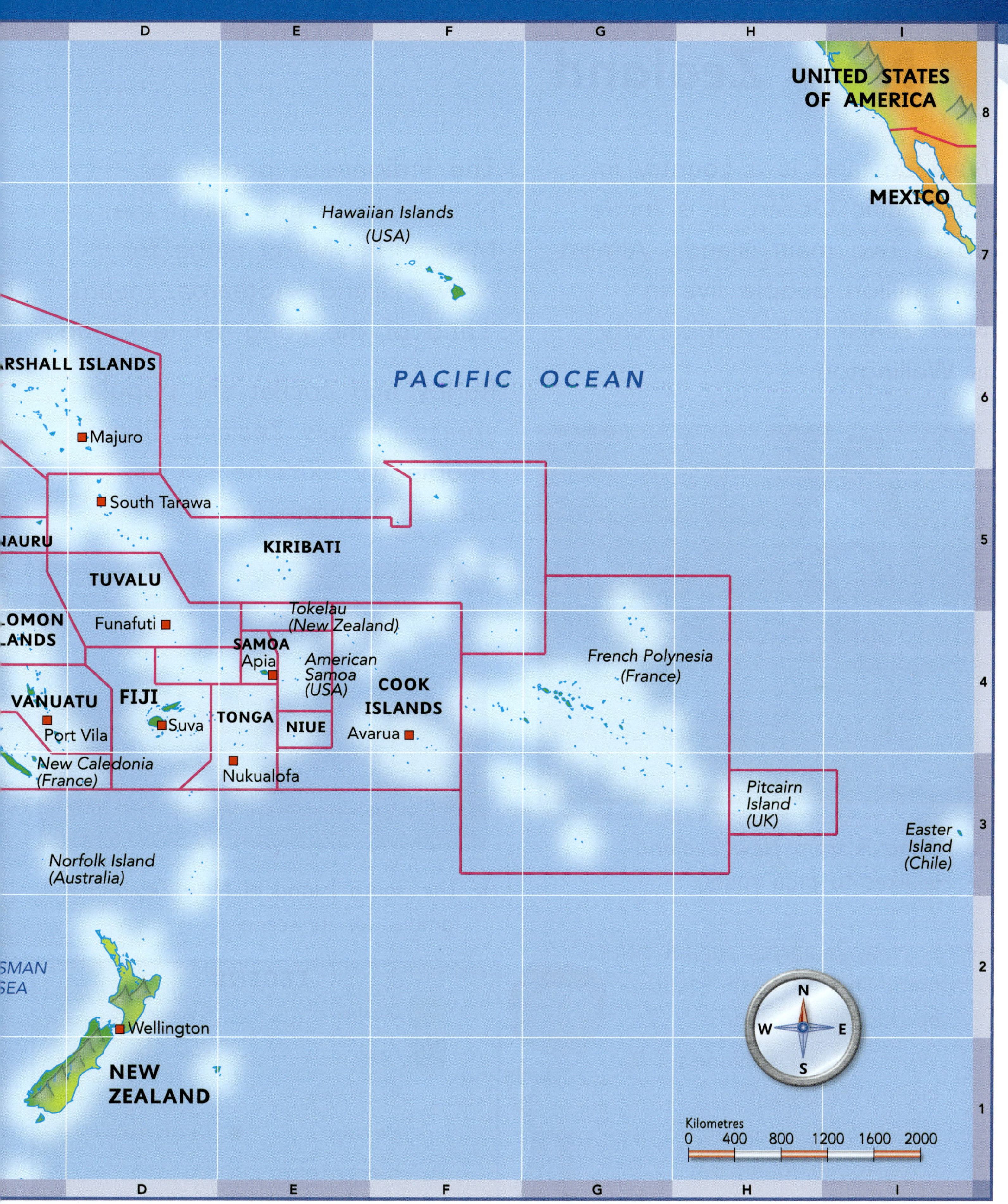
UNITED STATES OF AMERICA
MEXICO
Hawaiian Islands (USA)
PACIFIC OCEAN
MARSHALL ISLANDS
Majuro
South Tarawa
NAURU
KIRIBATI
TUVALU
Funafuti
SOLOMON ISLANDS
Tokelau (New Zealand)
SAMOA
Apia
American Samoa (USA)
COOK ISLANDS
French Polynesia (France)
VANUATU
FIJI
Suva
TONGA
NIUE
Avarua
Port Vila
New Caledonia (France)
Nukualofa
Pitcairn Island (UK)
Easter Island (Chile)
Norfolk Island (Australia)
TASMAN SEA
Wellington
NEW ZEALAND
N
W
E
S
Kilometres
0 400 800 1200 1600 2000
D E F G H I
8 7 6 5 4 3 2 1

New Zealand

New Zealand is a country in the Pacific Ocean. It is made up of two main islands. Almost five million people live in New Zealand. Its capital city is Wellington.

Rongo is from New Zealand. He likes to play rugby.

- Is New Zealand's capital city found in the North Island or the South Island?
- What is New Zealand's highest mountain?
- Which grid square is Lake Taupo found in?

The indigenous people of New Zealand are called the Māori. The Māori name for New Zealand, *Aotearoa*, means 'Land of the Long White Cloud'.

Rugby and cricket are popular sports in New Zealand. Some people try extreme sports, such as bungee jumping.

The South Island of New Zealand is famous for its scenery.

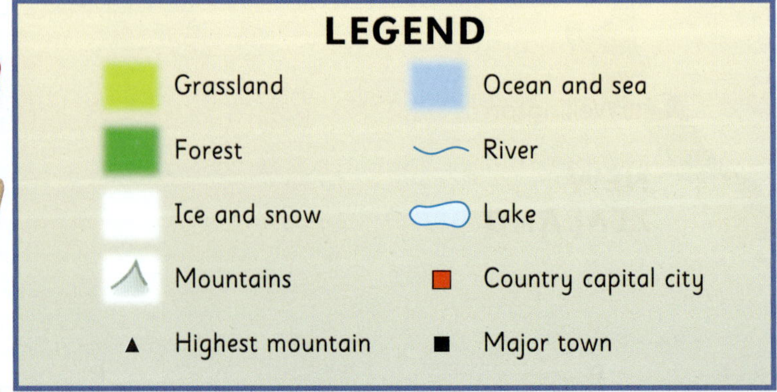

A
B
C
D
E
F
7
6
5
4
3
2
1
Bay of Islands
Auckland
Bay of Plenty
North Island
Hamilton
Waikato River
TASMAN SEA
Gisborne
New Plymouth
Lake Taupo
NEW ZEALAND
Cook Strait
Wellington
Greymouth
South Island
Aoraki/Mt Cook
SOUTHERN ALPS
Christchurch
PACIFIC OCEAN
Canterbury Plains
Waitaki River
Queenstown
Clutha River
N
W
E
S
Dunedin
Invercargill
Kilometres
0 50 100 150 200 250
Stewart Island

Asia

Asia is the largest continent. It is made up of 48 countries. More than four billion people live there. That is more people than in any other continent. The world's highest mountain, Mount Everest, is in Asia.

Hanako is from Japan. She is holding a paper crane.

- What is the capital city of Indonesia (F1)?
- Find the highest mountain in the world.
- What type of land covers the north of Russia?

D
E
F
G
H
I
ARCTIC OCEAN
RUSSIA
Lake Baikal
Astana
KAZAKHSTAN
Ulan Bator
MONGOLIA
BEKISTAN
Tashkent
Bishkek
KYRGYZSTAN
Dushanbe
TAJIKISTAN
Kabul
Islamabad
KISTAN
CHINA
Beijing
NORTH KOREA
Pyongyang
Seoul
SOUTH KOREA
JAPAN
Tokyo
PACIFIC OCEAN
HIMALAYAS
Mt Everest
New Delhi
NEPAL
Kathmandu
Thimphu
BHUTAN
Ganges River
BANGLADESH
Dhaka
Yangtze River
Taipei
TAIWAN
INDIA
MYANMAR
Naypyidaw
Yangon
LAOS
Hanoi
Vientiane
THAILAND
Bangkok
CAMBODIA
Phnom Penh
VIETNAM
Manila
PHILIPPINES
Colombo
SRI LANKA
Male
MALDIVES
BRUNEI
MALAYSIA
Kuala Lumpur
SINGAPORE
Jakarta
INDONESIA
Dili
TIMOR-LESTE
8
7
6
5
4
3
2
1

Europe

Europe is made up of 45 countries. The two smallest countries in the world are in Europe. These countries are Vatican City and Monaco. Some countries in Europe that are near the Arctic are very cold in winter. Other countries in the south of Europe have much warmer weather.

Freya is from Norway (E7). She is walking through the snow with her reindeer.

- Which river flows through Russia (H6)?
- Which country is west of Spain?
- Which grid square is London found in?

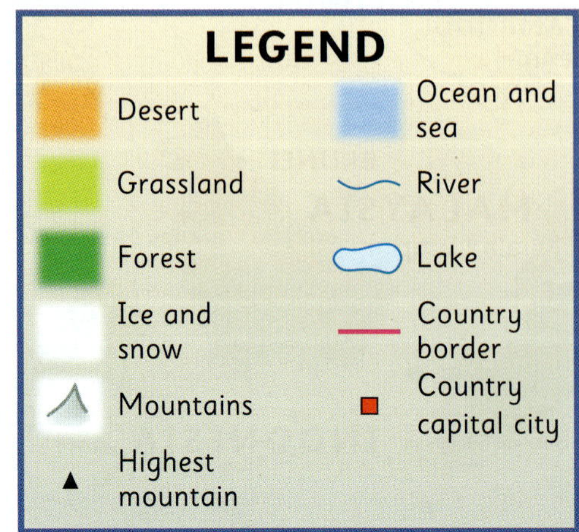

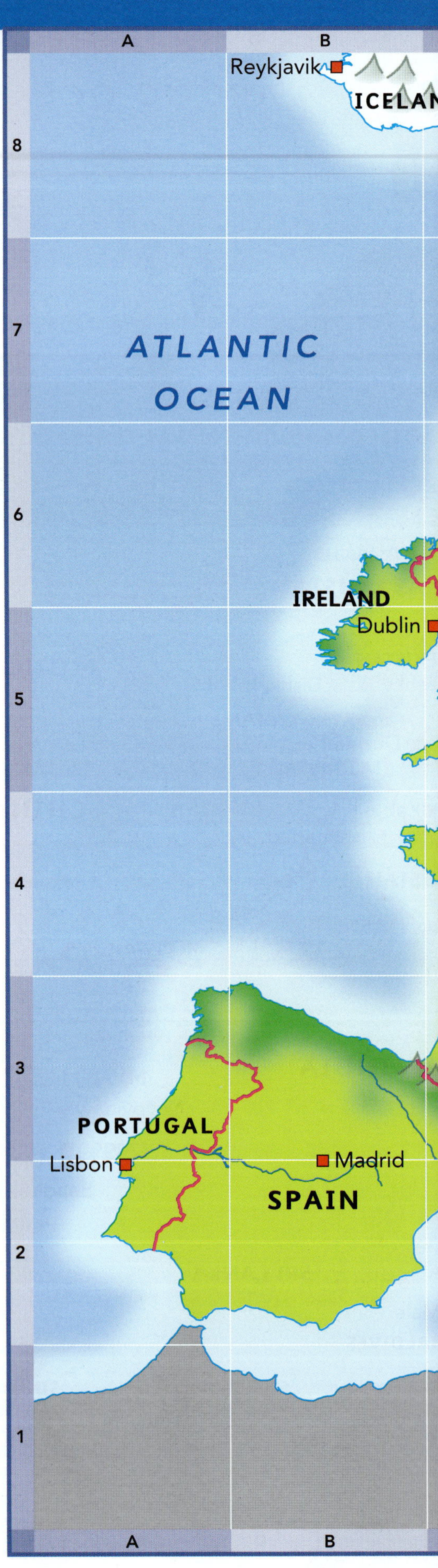

D
E
F
G
H
I
NORWEGIAN SEA
NORTH SEA
BALTIC SEA
NORWAY
SWEDEN
FINLAND
RUSSIA
Lake Ladoga
Volga River
Oslo
Stockholm
Helsinki
Tallinn
ESTONIA
LATVIA
Riga
Moscow
DENMARK
Copenhagen
UNITED KINGDOM
LITHUANIA
Vilnius
Minsk
BELARUS
NETHERLANDS
London
Amsterdam
Berlin
Warsaw
Brussels
BELGIUM
LUXEMBOURG
Paris
GERMANY
Rhine River
POLAND
Kyiv
UKRAINE
Prague
CZECH REPUBLIC
FRANCE
LIECHTENSTEIN
Vienna
Bratislava
SLOVAKIA
MOLDOVA
Kishinev
Bern
SWITZERLAND
AUSTRIA
Budapest
HUNGARY
THE ALPS
SLOVENIA
Ljubljana
Zagreb
ROMANIA
Mt Elbrus
CROATIA
MONACO
ANDORRA
SAN MARINO
BOSNIA and HERZEGOVINA
Sarajevo
Belgrade
SERBIA
Bucharest
BLACK SEA
ITALY
MONTENEGRO
Podgorica
KOSOVO
Pristina
BULGARIA
Sofia
VATICAN CITY
Rome
Skopje
Tirane
NORTH MACEDONIA
ALBANIA
GREECE
Athens
MEDITERRANEAN SEA
MALTA
N
W
E
S
Kilometres
0 150 300 450 600 750
8
7
6
5
4
3
2
1

Africa

Africa is the second biggest continent. There are 54 countries in Africa. A lot of north Africa is made up of desert. The Sahara Desert is the biggest desert in the world. The world's longest river is also in Africa. It is called the Nile River.

Omar is leading camels. He is from Egypt (F7).

- Which African country is furthest to the south?
- Which grid square is Africa's highest mountain found in?
- Which two oceans surround Africa?

D
E
F
G
H
I
Algiers
Tunis
TUNISIA
MEDITERRANEAN SEA
Tripoli
ALGERIA
LIBYA
Cairo
EGYPT
RED SEA
Sahara Desert
MALI
NIGER
Nile River
SUDAN
Khartoum
ERITREA
Asmara
Niamey
Ouagadougou
CHAD
N'Djamena
DJIBOUTI
Djibouti
BENIN
NIGERIA
TOGO
Abuja
Addis Ababa
Porto-Novo
SOUTH SUDAN
ETHIOPIA
CENTRAL AFRICAN REPUBLIC
Accra
Lome
CAMEROON
Bangui
Juba
Malabo
Yaounde
SOMALIA
EQUATORIAL GUINEA
Congo River
UGANDA
KENYA
Mogadishu
SAO TOME AND PRINCIPE
Sao Tome
Libreville
Kampala
Lake Victoria
GABON
DEMOCRATIC REPUBLIC OF CONGO
RWANDA
Kigali
Nairobi
CONGO
Mt Kilimanjaro
Brazzaville
Kinshasa
Bujumbura
BURUNDI
SEYCHELLES
Victoria
Dodoma
Luanda
TANZANIA
INDIAN OCEAN
COMOROS
Moroni
ANGOLA
MALAWI
ZAMBIA
Lilongwe
Lusaka
Zambezi River
Harare
MADAGASCAR
ZIMBABWE
Antananarivo
NAMIBIA
MAURITIUS
Port Louis
Windhoek
BOTSWANA
MOZAMBIQUE
Gaborone
Pretoria
Maputo
Mbabane
Bloemfontein
ESWATINI
Maseru
LESOTHO
SOUTH AFRICA
Cape Town
8
7
6
5
4
3
2
1

North America

There are 22 countries in North America. The largest country is Canada. There are many different indigenous peoples in North America.

The Pawnee people from Kansas in the United States and the Inuit people from the Arctic are two examples. Each group of indigenous people has different traditions.

Kuruk is wearing the costume of his indigenous tribe, the Pawnee.

Ahnah lives with her people, the Inuit, in northern Canada.

- What bay is found in D5?
- What is the capital city of the United States of America?
- Which mountain range runs through Canada and the United States of America?

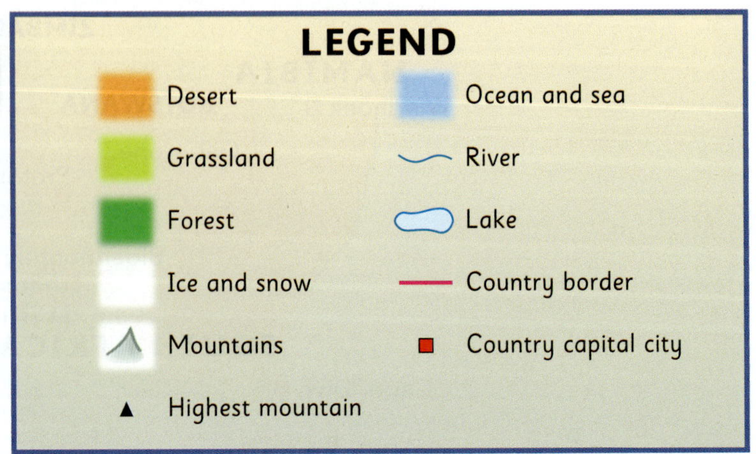

A B C D E F
7 6 5 4 3 2 1
ARCTIC OCEAN
Kalaallit Nunaat
(Denmark)
Alaska
(USA)
Denali
Baffin Island
Hudson
Bay
ROCKY
MOUNTAINS
CANADA
Newfoundland
Lake
Superior
Great
Basin
Missouri River
Ottawa
Colorado River
UNITED STATES
OF AMERICA
Washington DC
Mississippi River
PACIFIC
OCEAN
ATLANTIC
OCEAN
Gulf of Mexico
MEXICO
BAHAMAS
Nassau
Havana
CUBA
DOMINICAN REPUBLIC
Mexico City
Puerto
Rico
ANTIGUA
and
BARBUDA
HAITI
Port-au-Prince
Santo
Domingo
ST KITTS
and NEVIS
BELIZE
Belmopan
Kingston
JAMAICA
DOMINICA
ST LUCIA
BARBADOS
ST VINCENT
GRENADA
GUATEMALA
HONDURAS
Guatemala City
Tegucigalpa
CARIBBEAN
SEA
San Salvador
EL SALVADOR
Managua
NICARAGUA
San Jose
Panama City
COSTA RICA
PANAMA
N
W
E
S
Kilometres
0 400 800 1200 1600 2000

South America

There are 13 countries in South America. The biggest country is Brazil.

The Andes, on the west coast of South America, is the longest mountain range in the world.

The Amazon rainforest in South America is the largest rainforest in the world.

More than one-third of all plant and animal species in the world live in this rainforest.

Ana lives near Lake Titicaca (B5) at the foot of the Andes mountains.

Luis lives in the Amazon rainforest (C6).

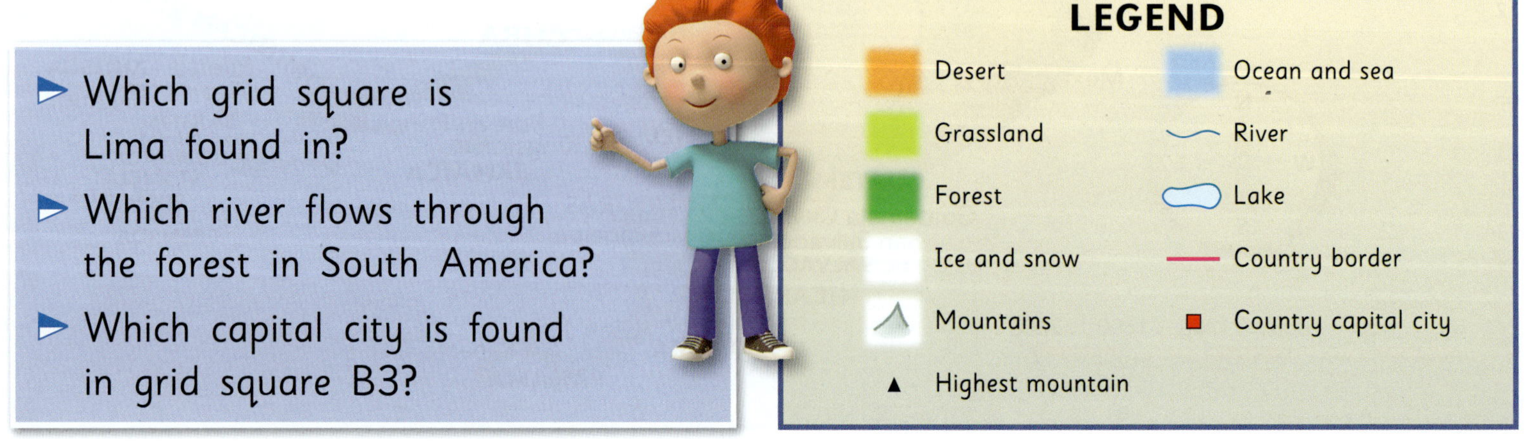

- Which grid square is Lima found in?
- Which river flows through the forest in South America?
- Which capital city is found in grid square B3?

A
B
C
D
E
F
7
6
5
4
3
2
1
CARIBBEAN SEA
Caracas
TRINIDAD AND TOBAGO
VENEZUELA
Orinoco River
Georgetown
GUYANA
Paramaribo
SURINAME
French Guiana
ATLANTIC OCEAN
Bogota
COLOMBIA
Quito
ECUADOR
ANDES MOUNTAINS
Amazon River
BRAZIL
PERU
Lima
Lake Titicaca
La Paz
BOLIVIA
Sucre
Brasilia
PARAGUAY
Asuncion
River
Parana
PACIFIC OCEAN
CHILE
ARGENTINA
URUGUAY
Santiago
Mt Aconcagua
Buenos Aires
Montevideo
ATLANTIC OCEAN
ANDES MOUNTAINS
Patagonian Desert
N
W
E
S
Kilometres
0
300
600
900
1200
1500

Antarctica

Antarctica is the coldest and driest continent in the world. Almost all of it is covered in thick ice. No trees or bushes grow in Antarctica. Seals and penguins spend time on the ice but then go back to sea. The only people who live in Antarctica are scientists. They go there for a short time to study the environment and wildlife.

This scientist is studying penguins.

- Which grid square is Mawson Station in?
- You can only travel in one direction from the South Pole. What is it?

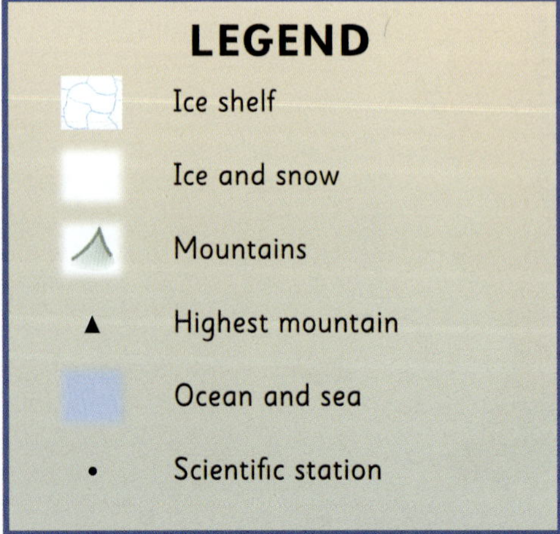

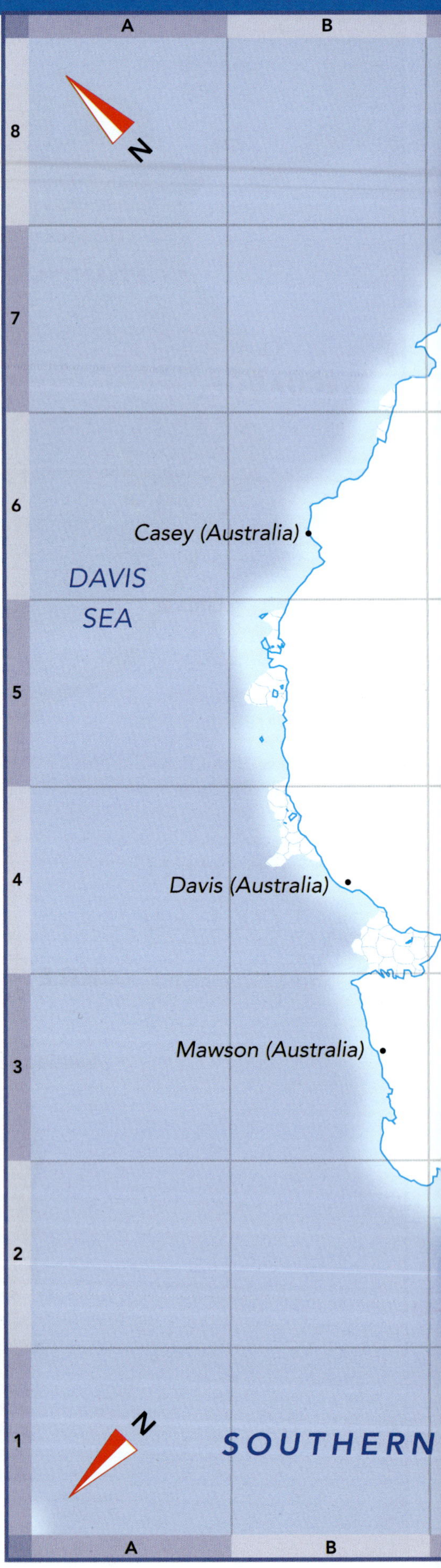

D
E
F
G
H
I
SOUTHERN OCEAN
ROSS SEA
Ross Island
Scott Base (NZ)
Roosevelt Island
Ross Ice Shelf
TRANSANTARCTIC MOUNTAINS
Greater Antarctica
Lesser Antarctica
Carney Island
AMUNDSEN SEA
Thurston Island
South Pole
Amundsen-Scott (USA)
Vinson Massif
Ronne Ice Shelf
Berkner Island
BELLINGSHAUSEN SEA
Antarctic Peninsula
Alexander Island
Larsen Ice Shelf
Drake Passage
Cape Horn
WEDDELL SEA
HAAKON VII SEA
CEAN
South Orkney Islands
SCOTIA SEA
South Sandwich Islands
Kilometres
0 250 500 750 1000 1250
N
N
8
7
6
5
4
3
2
1

Index

How to use this index:
place name — page number — grid reference

ACKNOWLEDGEMENTS

The author and the publisher wish to thank the following copyright holders for reproduction of their material.

Oceanwide Images/Gary Bell, front and back cover.

Photos: **123RF**, pages 29 (whale), 41 (dog), 51 top, 52 bottom, 68 left; **Agefotostock**/ Danita Delimont Stock, 96/KidStock/Blend Images, 42 bottom right/Robert Mabic, 53 bottom/Ronnie Kaufman/Larry Hirshowitz, 86; **Alamy**/Andrey Podkorytov, 55 (log)/Arco Images GmbH, 32–33/AXINITE, 74 right/Bill Bachman, 7 bottom left, 78 right, 7 top right/Blend Images, 57 top/David Noton Photography, 28–29/David Wall, 3 bottom, 7 bottom right, 84 right/Jeffery Drewitz, 29 (desert)/Jenny Matthews, 57 bottom/LEMAIRE Stphane/hem, 68 right/National Geographic Images, 42 bottom left/Penny Tweedie, 53 centre/Prisma by Dukas Presseagentur GmbH, 55 (stilt)/ScotStock, 60/Suzanne Long, 36; **Auscape** /Nicholas Birks, 31 top left/Glen Threlfo, 62/Tui De Roy, 35 (eagle); **Corbis**/Alan Towse; Ecoscene, 47; **Fairfax Images**, 58 bottom; **Getty Images/**Altrendo images, 43 bottom/Inmagineasia, 59 top left/ Per Breiehagen, 88/Yva Momatiukw/John Eastcott/Minden Pictures, 29 (zebras)/Harry Hook, 55 (mud)/Artic Images, 92 right/Auscape/UIG, 27 bottom/Bob Krist, 3 centre/Bob Krist, 3 centre, 94 left/Bruno Buongiorno Nardelli, 46–47/Dorling Kindersley/Ruth Jenkinson, 53 top/Frans Lemmens, 90/Fuse, 43 top right/Gary Vestal, 2 bottom/Gary Vestal, 3 bottom, 31 bottom/Image Source Pink, 57 centre/Inga Spence, 37 (orchard)/J. Parsons, 43 top left/J. Pat Carter, 92 left/Jason Edwards, 26 top left/Jef Meul/ Foto Natura, 33/Jeff Hunter, 70 left/John Carnemolla, 37 (dam)/Kevin Dodge, 42 top right/Lane Oatey, 48–49 (ping)/ Lily Chou, 74 left/London Scientific Films, 33/Luciano Candisani/Minden Pictures, 29 (rainforest)/M.M. Sweet, 34/Marty Melville/Stringer, 84 left/Mike Powles, 26 bottom left/Nancy Brown, 54/National geographic, 82/National Geographic/Stanley Breeden, 35 (possum)/Nik Wheeler, 56 bottom right/Owen Franken, 56 bottom left/Panoramic Images , 70 right/Planet Observer, 6/Ramin Talaie, 55 (glass)/Realfeel, 3 top, 4 top, 76/Rhys Pope, 27 top/Richard Gunn, 45 top/Rob Blakers, 72/Robin Smith, 45 bottom/Ross Barnett, 26 bottom centre/Rubberball/Erik Isakson, 48-49 (Zi)/Sara-Jane Cleland, 78 left/Stephan Dalton, 32/Stephan Dalton, 33/Ted Mead, 44/Tom Brakefield, 35 (leopard); **Imagefolk**/ Gerard Lacz, back cover (middle)/Minden Pictures, back cover (right)/Naturepl, 94 right; **Istockphoto**/jianying yin, 48–49/micheldenijs, 29 (bear); **Newspix**, 59 top right/Glenn Daniels, 37 (children)/Lyndon Mechielsen, 64 right, 46/Michael Chambers, 51 bottom; **Panos**/G.M.B Akash, 56 top; **Shutterstock**, 26, 27 bottom, 27 top, 30, 30, 31 top right, 35 (orang-utan), 35 (red panda), 37 (kangaroos), 40 (girl), 40 (lemon), 40 (melon), 40 (pebbles), 40 top left, 40–41 (background), 41 (guitar), 41 (lamp), 41 (maracas), 44–45 (paper), 48 background, 48–49, 49 background, 50 bottom, 50 top, 50–51, 51 centre, 52 top, 52–53, 54–55, 58 centre, 58 top, 58–59, 59 bottom, 64 left, 66, 7 top left.

Every effort has been made to trace the original source of copyright material contained in this book. The publisher will be pleased to hear from copyright holders to rectify any errors or omissions.

Oxford University Press is a department of the University of Oxford. It furthers the University's objective of excellence in research, scholarship, and education by publishing worldwide. Oxford is a registered trademark of Oxford University Press in the UK and in certain other countries.

Published in Australia by
Oxford University Press
Level 8, 737 Bourke Street, Docklands, Victoria 3008, Australia

First published 2013
Revised edition 2018
Reprinted 2018, 2019, 2020, 2021, 2022, 2023, 2024, 2025 (twice)

National Library of Australia Cataloguing-in-Publication entry
Title: Oxford atlas+ for Australian schools F-2 / Oxford University Press.
Edition: Revised edition
ISBN: 978 0 19 031076 9 (paperback)
Notes: Includes index
Target Audience: For primary school age
Subjects: Atlases, Australian–Juvenile literature

'Our amazing senses' (pp. 40–41), 'Families' (pp. 50–51), 'Remembering the past' (pp. 52–53), 'Celebrations around Australia' (pp. 50–59) written by Rachel Kennedy
Edited by Emma Short and Katrina Spencer
Original text design by Christina Neri
Typeset and designed by Rose Keevins, Eggplant Communication and Oxford University Press
Cartography and map index by MAPgraphics Pty Ltd, Brisbane
Printed in Hong Kong by Sheck Wah Tong Printing Press Ltd.

Aboriginal and Torres Strait Islander people are warned that this book may contain images of deceased people.

AUSTRALIA AND THE PACIFIC

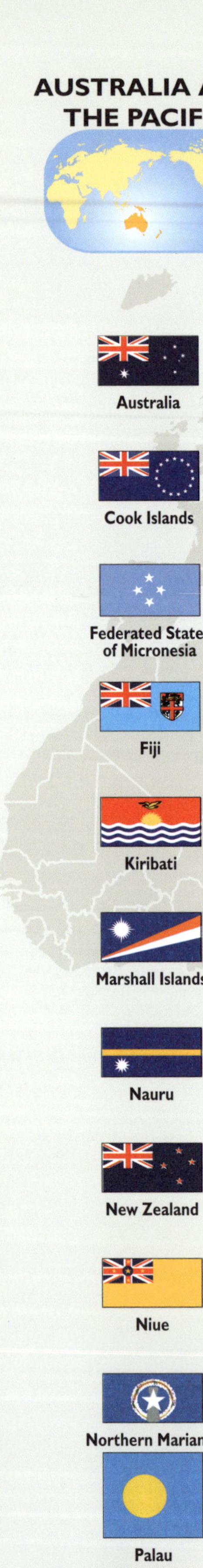
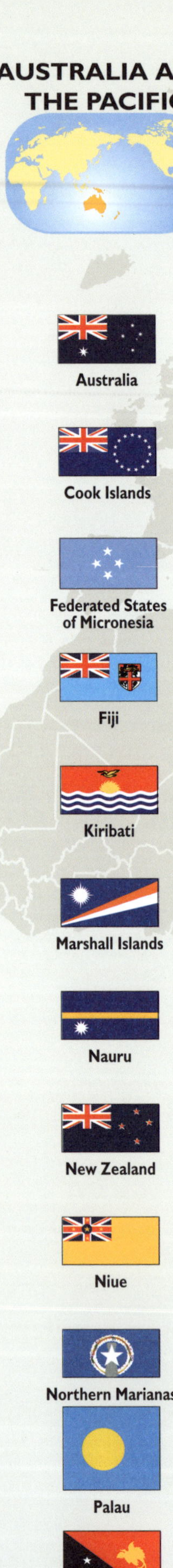
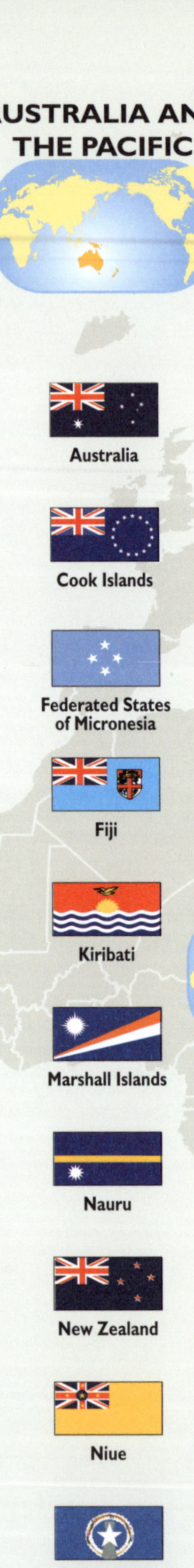

Australia
Cook Islands
Federated States of Micronesia
Fiji
Kiribati
Marshall Islands
Nauru
New Zealand
Niue
Northern Marianas
Palau
Papua New Guinea
Samoa
Solomon Islands
Tonga
Tuvalu
Vanuatu

ASIA

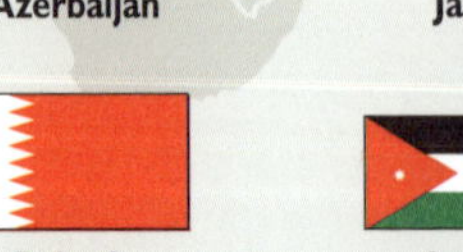
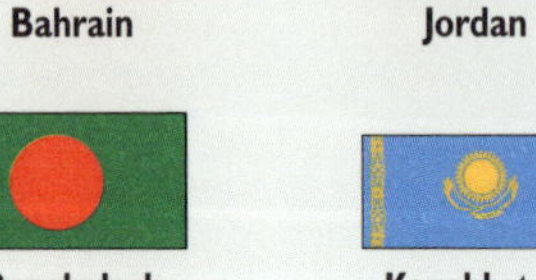

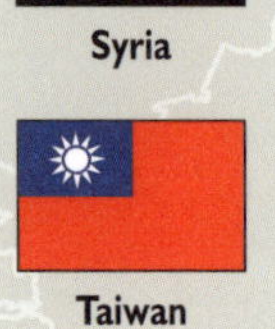

Afghanistan
Armenia
Azerbaijan
Bahrain
Bangladesh
Bhutan
Brunei
Cambodia
China
Cyprus
Georgia
India
Indonesia
Iran
Iraq
Israel
Japan
Jordan
Kazakhstan
Kuwait
Kyrgyzstan
Laos
Lebanon
Malaysia
Maldives
Mongolia
Myanmar
Nepal
North Korea
Oman
Pakistan
Philippines
Qatar
Saudi Arabia
Singapore
South Korea
Sri Lanka
Syria
Taiwan
Tajikistan
Thailand
Timor-Leste
Turkey
Turkmenistan
United Arab Emirates
Uzbekistan
Vietnam
Yemen

EUROPE

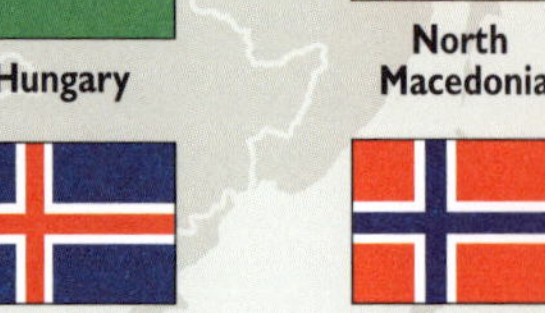
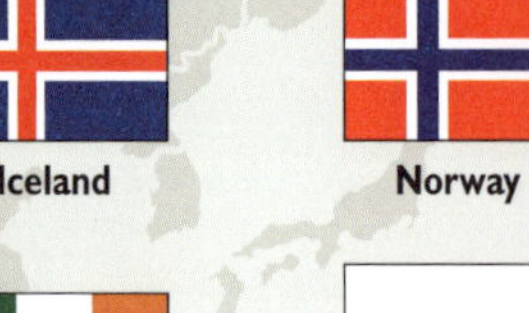

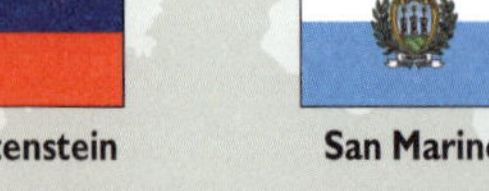

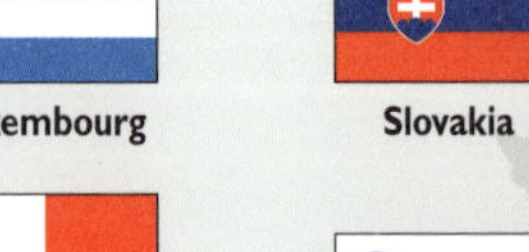

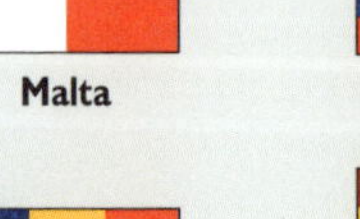

Albania
Andorra
Austria
Belarus
Belgium
Bosnia and Herzegovina
Bulgaria
Croatia
Czech Republic
Denmark
Estonia
Finland
France
Germany
Greece
Hungary
Iceland
Ireland
Italy
Kosovo
Latvia
Liechtenstein
Lithuania
Luxembourg
Malta
Moldova
Monaco
Montenegro
Netherlands
North Macedonia
Norway
Poland
Portugal
Romania
Russia
San Marino
Serbia
Slovakia
Slovenia
Spain